Voice Leading

Voice Leading

The Science behind a Musical Art

David Huron

The MIT Press
Cambridge, Massachusetts
London, England

© 2016 Massachusetts Institute of Technology

All rights reserved. No part of this book may be reproduced in any form by any electronic or mechanical means (including photocopying, recording, or information storage and retrieval) without permission in writing from the publisher.

MIT Press books may be purchased at special quantity discounts for business or sales promotional use. For information, please email special_sales@mitpress.mit.edu or write to Special Sales Department, The MIT Press, 1 Rogers Street, Cambridge, MA 02142.

This book was set in Stone Sans and Stone Serif by Toppan Best-set Premedia Limited. Printed and bound in the United States of America.

Library of Congress Cataloging-in-Publication Data

Names: Huron, David Brian.
Title: Voice leading : the science behind a musical art / David Huron.
Description: Cambridge, MA : MIT Press, [2016] | Includes bibliographical references and index.
Identifiers: LCCN 2016004808 | ISBN 9780262034852 (hardcover : alk. paper)
Subjects: LCSH: Harmony. | Music--Psychological aspects. | Music perception.
Classification: LCC ML3836 .H87 2016 | DDC 781.4/2--dc23 LC record available at http://lccn.loc.gov/2016004808

10 9 8 7 6 5 4 3 2 1

Contents

Preface   vii

1 Introduction   1
2 The Canon   9
3 Sources and Images   13
4 Principles of Image Formation   27
5 Auditory Masking   41
6 Connecting the Dots   63
7 Preference Rules   87
8 Types of Part-Writing   97
9 Embellishing Tones   121
10 The Feeling of Leading   129
11 Chordal-Tone Doubling   149
12 Direct Intervals Revisited   157
13 Hierarchical Streams   163
14 Scene Setting   173
15 The Cultural Connection   185
16 Ear Teasers   195
17 Conclusion   207

Afterword   219
Acknowledgments   221
Notes   223
References   237
Index   255

Preface

In normal conversation, people take turns so that only one person speaks at a time. Listeners are often baffled when two people speak simultaneously. In music, by contrast, polite turn taking is the exception: musicians seem perfectly happy to all "talk" at once. Concurrent sounds raise special perceptual challenges. Listening to two melodies at the same time can be as confusing as listening to two simultaneous conversations—unless the melodies are carefully constructed and coordinated.

Over the centuries, musicians have formulated various rules or recommendations based on their practical experience in arranging multiple concurrent sounds. Over time, the various rules have coalesced into different canons of practice. Part-writing rules differ from one style to another, so writing species counterpoint differs from writing madrigals, harmonizing chorale melodies, composing barbershop quartets, or arranging for jazz combo. Despite the many style-related differences, a coherent core informs all types of part-writing.

This book has two complementary aims. The first aim is to provide a scientific explanation for the core part-writing rules. The book assumes that readers will already have some facility with part-writing. Explanations will be offered for both part-writing and the larger umbrella concept of *voice leading*. By conveying a better understanding of the pertinent perceptual research, I also hope to help musicians improve their part-writing skills. To this end, the concluding chapter offers detailed advice for advanced part-writing.

The second aim of this book is to provide a science-based account of musical texture. We will see that voice leading ultimately opens doors to much broader issues of sonic organization—what we will refer to as musical *scene setting*. Coordinating concurrent musical sounds is not merely a matter of independent part-writing; it also entails integrating musical resources so they cohere perceptually. Both tasks are central to composing, arranging, and orchestration. And both tasks are applicable to all kinds of music—not just classical music. The same organizational issues are faced by all music-makers, including electroacoustic musicians and abstract sound designers.

In addressing these two aims, this book will summarize one of the most successful and exciting areas in auditory research. Over the past sixty years, scientific scholarship has helped to explain how humans hear and how listeners make sense of the world of sound. Inspired by this work, my collaborators and I have carried out dozens of empirical studies applying this research to music. The resulting story has proved enormously gratifying. Unfortunately, the published scientific accounts often assume a sophisticated technical background and thus present a challenge for many interested musicians and music lovers. In writing this book, I have sought to make this research accessible without oversimplifying it.

In the end, readers should come away with a solid understanding of the perceptual and cognitive foundations behind a centuries-old musical practice, and be better prepared to create complex new sonic textures suitable for more modern styles. Readers will learn that, far from being arbitrary, the core rules of part-writing arise from an underlying logic that is remarkably clear, perhaps even beautiful.

It bears emphasizing that our task is descriptive rather than prescriptive. That is, my intention is to describe some musically pertinent aspects of human hearing, not to dictate or advocate adherence to some aesthetic ideal. The purpose is to provide musicians with a better understanding of the tools they use, not to tell musicians what goals they should pursue.

As our story unfolds, it is important for readers not to assume that the auditory principles described here are innate or universal. This impression will be allowed to fester until chapter 15, when I review research suggesting that many of the perceptual foundations of voice leading are themselves learned from exposure to an acoustic environment. We will see scientific evidence that what you listen to influences how you hear. The research suggests that a listener's musical culture and individual listening habits shape the manner in which that person hears musical parts and textures. In the same way that languages change over the course of history, voice leading is a work in progress rather than a fixed practice.

This book has accompanying sound demonstrations that may be found on the web. The website also contains supplementary teaching materials. A search for "voice leading demonstrations" should lead readers to the site.

1 Introduction

Knowledge of voice leading has been widely viewed as one of the essential foundations in the training of musicians. *Voice leading* has been variously defined, but one simple definition is that it is the art of combining concurrent musical lines or melodies. The art of voice leading was developed by early Renaissance polyphonists in Germany, France, and the Netherlands. In a relatively short time, the practice became codified into a set of recommendations (do's and don'ts). The codified practice has evolved over hundreds of years, reflecting new insights as well as changes in musical styles and tastes. In one form or another, some sort of canon of rules has been taught to successive generations of musicians. In recent times, a single practice has dominated the conservatory curriculum: most musicians have been taught Baroque chorale-style part-writing. Mastery of part-writing requires more than simply adhering to a bunch of do's and don'ts; nevertheless, the rules provide a helpful starting point.

Like any other subject, the subject of voice leading has had its share of disagreements. It has been interpreted differently by different scholars and in different periods. Musicians have disagreed about its purpose and nature, and whether voice leading pertains only to certain musical styles. Most of all, music theorists have disagreed about which specific rules do or don't belong in some voice-leading canon and have quibbled over the precise wording of various rules. Many music students, especially students of jazz and popular music, have rightly questioned whether the traditional practice of voice leading serves any useful purpose in their training. As we'll see, even when specific part-writing practices are irrelevant, voice leading offers some deep musical insights pertinent to all musical styles.

I first encountered voice leading as a young teenager. I was not an especially cooperative music student and constantly pestered my teachers with questions. "But why?" I would ask. "Why avoid parallel fifths?" "What's wrong with crossing parts?" A couple of rules seemed to make intuitive sense, such as the rules limiting the ranges of the voices. However, most of the rules seemed arbitrary, and the idea of having rules went against my sense that composers could create music however they wanted.

I still believe that composers are free to do whatever they want, but I now know that there are consequences to any given course of action, and it is often useful to be aware of the repercussions before doing something. In the case of voice leading, we now have a better idea of the perceptual consequences of different ways of arranging sounds. The purpose of this book is *not* to tell musicians how to create music. Instead, this book presents a sort of road map that describes what happens when a musician chooses one path rather than another.

Voice leading has been typically taught as a set of rules. When considering any rule, we can ask three basic questions:

1. What goal is served by following the rule?
2. Is the goal worthwhile?
3. Is the rule an effective way of achieving the goal?

No composer should feel obliged to follow any rule unless he or she agrees that the rule provides an efficient path toward a worthwhile destination. Consider, for example, the traffic rule: *a red light means stop and a green light means go.* The goal of this rule is to ensure safety while facilitating traffic flow. It does this by establishing a clear right-of-way. *Is the goal worthwhile?* Certainly. *Is the rule an effective way of achieving the goal?* For most people, yes—although we might quibble with it. The rule is useless for blind pedestrians and baffling for the small number of drivers who have red/green color blindness. Since the confusion of red and green is the most common form of color blindness, a more effective rule might have contrasted red with, say, blue.

The same three questions can be asked when encountering a rule such as the following version of the *hidden* or *direct octaves* rule in voice leading:

When outer voices approach an octave by similar motion, be sure that the uppermost part moves by diatonic step.

What purpose is served by following this seemingly peculiar rule? Is the goal musically worthwhile? And can we improve the rule so that it would better facilitate achieving the goal? This book addresses these questions for common voice-leading rules. We will see that there is a cogent logic to the traditional rules and that they all contribute to a particular goal: *voice leading is a codified practice that helps musicians craft simultaneous musical lines so that each line retains its perceptual independence for an enculturated listener.* We will also see that perceptual independence can cause the brain to experience a particular type of pleasure. In other words, when part-writing conforms to the voice-leading canon, enculturated listeners typically find that the result sounds pleasant.

Of course, this doesn't mean that composers must or should follow any of the rules of voice leading. There is no law that all musical works must evoke pleasure in listeners. And even if a composer wants to create a pleasurable effect, the specific pleasure evoked by voice leading is not the only form of pleasure that can be experienced through sound. Moreover, even if a composer aims to evoke the pleasure afforded by voice leading, he or she might also want to pursue other concurrent goals that might sometimes conflict with the goals of voice leading. As we will see, pursuing a mix of compositional goals can lead to compromises that require modifications to a given voice-leading practice.

Over the centuries, music theorists have written extensively about the practice of voice leading, and many valuable insights have accumulated.[1] Prior to the twentieth century, theorists tended to view the voice-leading canon as a set of fixed and inviolable universals. (Musicians *must* obey these rules—or else!) By the mid-twentieth century, scholars began to regard the voice-leading canon as descriptive of a particular historical musical convention. For some scholars, the rules themselves were arbitrary conventions that define a style, like choosing to wear blue jeans and a T-shirt as opposed to a tuxedo. We now have a more nuanced understanding of the voice-leading canon. Although voice leading represents just one of countless considerations when creating music, the rules also provide important insights about auditory processing and hold general lessons for composers apart from whether they choose to follow the practice.

Traditionally the study of harmony entails two parts: (1) advice regarding harmonic organization and (2) recommendations regarding part-writing. The first part pertains to the preferred progression of chords, including the overall harmonic plan of a work, the choice and placement of cadences, and the moment-to-moment succession of individual chords. The second part specifies the manner in which individual parts or voices move from tone to tone in successive sonorities. The English term *voice leading* is a literal translation of the German *Stimmführung* and refers to the direction of movement for tones within a single part or voice. Various theorists have long suggested that the principal purpose of voice leading is to create perceptually independent musical lines. This book describes a body of scientific research that endorses this view.

In this book, our focus is principally on the practice of voice leading. No attempt will be made to account for the traditional advice regarding harmonic organization. The goal is to explain voice-leading practice using perceptual principles—predominantly principles associated with the theory of *auditory scene analysis*.[2] As we will see, this approach provides an especially strong account of voice-leading practice.

Plan

This book is organized according to the following plan. Chapter 2 briefly reviews the traditional rules of voice leading as formulated in the late Baroque period. Chapter 3 provides a general introduction to auditory perception. Beginning with chapter 4, various principles of auditory perception are described and their musical implications identified. In particular, chapter 4 discusses why certain kinds of timbres are more suited to creating musical lines than others. Chapter 5 discusses how sounds interfere with each other ("masking") and suggests why musicians in nearly every culture tend to place the main melody in the highest voice. Chapter 6 considers how successive pitches can give rise to a sense of *line*.

Six auditory principles described in chapters 4, 5 and 6 provide the basis for a detailed explanation of the core rules of voice leading given in chapter 7. In presenting this explanatory account, several novel rules will arise that are not found in past writings on voice leading. These novel rules can be regarded as theory-generated predictions. If the perceptual theory is correct and if past composers regarded the goal of perceptually independent musical lines as worthwhile, then we might expect composers to write in a manner consistent with these novel rules. Indeed, that's what we find: music written by past composers proves to be consistent with the new rules predicted by the perceptual theory.

Chapter 8 presents four additional perceptual principles that are occasionally linked to the practice of voice leading. In essence, these additional principles have been treated as compositional options that shape music-making in perceptually unique ways. Some composers have written in accordance with most or all of these principles, and others have not. Depending on which auxiliary principles are followed, we will see that alternative voice-leading systems arise. For example, the mix of principles used leads to the distinction between homophonic and polyphonic part-writing and also to such unique styles as close harmony (such as the voice leading found in barbershop quartets).

Chapter 9 discusses embellishing tones, such as passing tones, suspensions, and appoggiaturas. The chapter focuses on how these embellishing tones can contribute to the perceptual independence of concurrent voices. Chapter 10 continues with a discussion of the psychological feeling of tones "leading" somewhere. Research on melodic expectation is summarized, and we discuss how feelings of yearning, tension, and resolution arise. Chapter 10 also makes a belated distinction between part-writing and voice leading. We'll see examples of "correct" part-writing where the music nevertheless sounds like it is meandering aimlessly: voice leading adds the sense of tending, striving, or momentum that contributes to the feeling of music "going somewhere."

Chapter 11 addresses the arcane rules of chordal-tone doubling. For more than two centuries, there have been competing sets of doubling rules and an ongoing controversy about which rules are "correct." This chapter reviews the most careful empirical studies that have addressed this topic. We'll see that the competing sets of rules predict virtually identical outcomes, and so it matters little which rule set is used. Nevertheless, one set of rules is expressed in a way that is more consistent with the underlying auditory principles.

Chapter 12 examines the phenomenon of "direct" or "hidden" octaves. Perceptual experiments testing this traditional rule will lead us to question whether listeners hear nominally four-part harmony as truly evoking four independent lines. This observation leads to chapter 13, which discusses hierarchical streams; the chapter introduces *scene analysis trees* as a graphical analytic tool for better understanding different kinds of musical textures.

Chapter 14 discusses musical *scene setting*, such as different tune-and-accompaniment treatments. In addition, the chapter addresses the question as to why so many introductory music theory courses focus on Baroque voice-leading rules to the virtual exclusion of other types of part-writing. Although most music-making bears little resemblance to Baroque-style four-part chorale writing, we will see that there are excellent reasons why this particular practice has occupied the center of the core theory curriculum for so long. For example, we will see how the perceptual principles underlying voice leading provide an important entry point for understanding any musical texture—no matter what the style, culture, or genre of music-making.

Chapter 15 considers the role of learning in how listeners experience different musical textures. We will see how changing musical cultures influence hearing.

Chapter 16 addresses the "what for" question. It describes a series of perceptual experiments that investigate the pleasingness of different musical organizations and offers a psychological account for why many listeners experience conventional voice leading as pleasing.

The main ideas of the book are briefly summarized in chapter 17. This concluding chapter also presents a refined and expanded canon of part-writing advice for advanced music students. The chapter closes by offering concrete tips for performers and conductors that arise from the science of voice leading. For those readers who teach voice leading and who would like to incorporate aspects of the pertinent research into their classes, an afterword is provided that offers some essential pedagogical advice.

The purpose of this book is not somehow to justify or defend traditional voice leading. Nor is the intent to restrict in any way the creative enterprise of musical composition. If a composer chooses a particular goal explicitly or implicitly, there are frequently consequences that may constrain the music-making in such a way as to make the goal achievable. The goals that composers pursue can include social goals; political goals; religious or historical goals; formal, perceptual, cognitive, emotional goals; or a host of other aims. Moreover, different composers prioritize their goals differently. After reading this book, readers should not conclude that voice leading is the most important musical goal or that other goals are somehow less important or irrelevant to music-making. This book merely identifies and clarifies some of the cognitive and cultural factors that bear on the perception of multiple musical lines. Some of these factors have had a profound and demonstrable influence on music-making around the world and across documented history. They are likely to continue influencing music well into the future.

Finally, it's worth noting that some of the points demonstrated in this book will seem obvious or even trivial to experienced musicians. Rigorous research can sometimes lead to unexpected new insights, but rigorous analysis can just as often take a very long path to arrive at an answer that no one ever doubted. Attempts to explain common sense can often seem ridiculous, and this is also true when explaining musical common sense. For readers who can tolerate a few geeky descriptions of the obvious, your reward will be some unexpected new insights.

2 The Canon

For more than twenty years, Anna Magdalena Bach (Johann Sebastian's second wife) kept a handsome gold-edged hard-covered notebook containing short handwritten compositions. Some were composed by her husband, and a few musical ditties were offered by visitors to the Bach household. In addition, the notebook contained a smattering of harmony exercises done by the Bach children. Near the end of the book is a text written in the juvenile scrawl of a very young Johann Christoph Friedrich Bach: it lists the main rules of voice leading as practiced in seventeenth-century Germany.[1]

Very few musicians have had the opportunity to learn voice leading from a master like J. S. Bach. Most music students today learn voice leading from a single textbook with the help of one or more teachers. They typically learn a version of Baroque chorale-style practice, although some students will be introduced to species counterpoint as well. Even in the case of Baroque style, unless you spend time looking at different textbooks, it is easy to form the impression that the rules of voice leading are absolutely fixed. In fact, there is considerable variation from text to text even today. The variation in voice-leading rules is especially obvious in books published before about 1970. The more people communicate with each other, the greater the tendency is to standardize things. For example, if you look at English-language newspapers before 1900, you'll be amazed at the range of English spellings. It wasn't just that British and American spellings differed. There were also different spellings in Boston and Philadelphia. Even newspapers in the same city sometimes used different spellings. As people communicated more, spellings became more and more standardized. Of course there is no absolutely "correct" spelling. Whether we spell it "neighbor" or "neighbour"—or even "nabur"—is a matter of convention. The "correct" spelling is the one everyone else uses.

This same process of standardization influenced the canon of voice-leading rules. In the 1960s, the number of university and conservatory students in Western countries expanded dramatically. Like different English spellings, differences between textbooks became something of an embarrassment. By about 1970, commercial textbook publishers had become more sensitive to teacher feedback that this or that element of a textbook was "unconventional." As a result, textbooks rapidly converged so as to conform to a majority view. We'd like to think that only the "right" rules survived, but that's probably too optimistic. There's more to voice leading than any individual musician has learned. Despite the standardization of the past decades, some variation in views remains. Even today, both theoretical books and textbooks on voice leading differ in detail. Some texts identify a handful of voice-leading rules, while others define dozens of rules and their exceptions. Some methods emphasize a historical approach, whereas others offer a more analytical presentation.[2] In later chapters, we'll see how even popular recent textbooks disagree about certain rules.

Rules of Voice Leading Reviewed

Despite the variety, a core group of rules appears in nearly every pedagogical work on voice leading. These core voice-leading rules describe Baroque chorale style:

1. *Compass rule.* Texts typically begin by specifying the pitch range for part-writing. Students should write in the region between F2 (at the bottom of the bass staff) and G5 (at the top of the treble staff). This range corresponds to the combined pitch range for typical male and female voices, but it is noteworthy that students are encouraged to write in this region even when writing purely instrumental music.
2. *Textural density rule.* Part-writing typically involves three or more concurrent "parts" or "voices." The most common part-writing employs four voices set in overlapping pitch ranges or tessituras: soprano, alto, tenor, and bass.
3. *Chord spacing rule.* In the spacing of chordal tones, no more than an octave should separate the soprano and alto voices. Similarly, no more than an octave should separate the alto and tenor voices. In the case of the bass and tenor voices, however, no restriction is placed on the distance separating them.

4. *Unisons rule.* Two voices should not share the same concurrent pitch. (An exception is made for chords that end a phrase or a musical work.)

5. *Common tone rule.* Pitch classes common to consecutive chords should be retained in the same voice or part.

6. *Nearest tone rule.* If a part cannot retain the same pitch in the next sonority, then the part should move to the nearest available pitch.

7. *Step motion rule.* When a part must change pitch, the preferred pitch motion is by diatonic step. Sometimes this rule is expressed in reverse, that is:

8. *Avoid leaps rule.* Large melodic intervals should be avoided.

9. *Part crossing rule.* Parts should not cross in pitch.

10. *Part overlap rule.* No part should move to a pitch higher than the immediately preceding pitch in an ostensibly higher part. Similarly, no part should move to a pitch lower than the immediately preceding pitch in an ostensibly lower part.

11. *Parallel perfects rule.* No two voices should move in parallel perfect octaves, perfect fifths, or their compound equivalents (e.g., unisons, fifteenths, twelfths). Many theorists use a more stringent version of this rule:

12. *Consecutive perfects rule.* No two voices should form consecutive unisons, octaves, fifteenths (or any combination thereof)—whether or not the parts move in parallel. Also, no two voices should form consecutive fifths, twelfths, nineteenths (or any combination thereof).

13. *Direct (or hidden or exposed) octaves (and fifths) rule.* Unisons, perfect fifths, octaves, twelfths, and so on should not be approached by similar motion between two parts unless at least one of the parts moves by diatonic step. Most theorists restrict this rule to the case of approaching an octave (fifteenth, etc.), whereas some theorists also extend this injunction to approaching perfect fifths (twelfths, etc.). Most theorists restrict this rule to voice leading involving the bass and soprano voices; in other cases, direct intervals between any pair of voices are forbidden. Many statements of this rule add the further stipulation that the step motion must occur in the soprano.

14. *Tendency doubling rule.* When duplicating chordal tones in some sonority, doubling the leading tone should be avoided. Also avoid doubling tones that are foreign to the key (i.e., chromatic pitches).

15. *False (or cross-) relation rule.* Avoid two successive sonorities where a chromatically altered pitch appears in one voice but the unaltered pitch

appears in the preceding or following sonority in another voice. (This is also sometimes called an English dissonance.)

16. *Augmented intervals rule.* Avoid augmented melodic intervals.

Depending on how you were taught, you will find this or that rule missing or differently phrased in the list. Music teachers in Italy, German, United States, France, Sweden, and Britain have long quibbled over various details. For reasons that will become apparent in chapter 11, I have omitted most of the traditional rules related to chordal-tone doubling. For example, the rule recommending doubling the root of the chord and the rule to avoid doubling the third are missing. Also missing is the less common rule recommending doubling of the tonic, dominant, and subdominant pitches.

Over the centuries, a number of music theorists have suggested that voice leading is the art of creating independent parts or voices. For the remainder of this book, we will take this suggestion literally and link voice-leading practice to scientific research concerning the perception of independent concurrent sounds. We will see that this apparently ragtag collection of rules forms a coherent canon with a deep unifying logic. We will also see how voice-leading practice points to much bigger issues of musical organization that transcend four-part chorale-style writing to impact all forms of music-making.

3 Sources and Images

Sounds are caused by mechanical movements that disturb the surrounding air. The sound may be caused by a brief burst of energy (such as an object falling on the ground) or a more sustained energy source (such as two objects rubbing against each other).

What we call "a sound" is actually pretty complicated. Even a simple sound, like a single tone played on a flute, is really a collection of component frequencies called *partials*. A partial is an elementary "part" of a sound that cannot be further broken apart. The discovery that nearly all sounds consist of such "subsounds" is the single most important acoustical discovery in history. Yet the discovery was made over twenty-five hundred years ago in both ancient China and ancient Greece. Each partial has a unique frequency (or rate) of vibration. Partials exist because most vibrating objects can vibrate in more than one way at the same time. If you strike a cube of gelatin with a spoon, for example, it won't simply rock back and forth; rather, it will wobble and jiggle in a complicated dance. That is, the gelatin will vibrate in several different ways simultaneously—left to right, back and forth, up and down, twisting clockwise then counterclockwise, and so on. Technically, each manner of vibration is referred to as a *mode*. Each mode has its own distinct frequency (or rate) of vibration for a given vibrating object.

The unit of frequency is the *hertz* (*Hz*). One hertz is one cycle of movement per second. When you walk, your head bobs up and down about two times each second: its frequency of movement is about 2 Hz.[1] Your head bobbing generates a sound, but its frequency is too low for humans to hear. A wobbling cube of gelatin also produces frequencies that are too low for humans (although elephants can probably hear it). When you knock on a door, the energy you impart will cause the door to wobble much like the cube of gelatin, although the movement will be much smaller and much faster than the gelatin cube. Because the rate of vibration is so much faster, knocking on the door will generate a sound humans can hear. A given vibrating door might produce a sound consisting of six main partials whose frequencies are (say) 48 Hz, 119 Hz, 142 Hz, 703 Hz, 972 Hz, and 1,488 Hz. Each of these frequencies arises from a different mode of vibration. That is, each partial arises from a different way in which the door bends and flexes in response to being struck. When a sound source vibrates in only one mode it produces a single partial and the resulting sound is referred to as a *pure tone*. When a source vibrates in two or more modes it produces two or more partials and the resulting sound is referred to as a *complex tone*. Nearly all sounds heard in the real world are complex. For most complex sounds, all of the partials will appear at the same time; however, some partials are slower to get started, and some may die away sooner than others.

Depending on the sound source, the partials may be simple multiples of some basic frequency: that is, they may form a series of frequencies that are two, three, four … times some starting value. An example of such a series is 101 Hz, 202 Hz, 303 Hz, and 404 Hz. Each successive frequency is a multiple of 101, which is called the *fundamental* frequency for this set of partials. When partials have frequencies that are related by simple integer multiples, they are referred to as *harmonics*. All harmonics are partials, but not all partials are harmonics. Knocking on a door produces many partials but no harmonics.

Acousticians have shown that there are different classes of vibrating objects. For example, membranes (like balloons, bubbles, and drums) have distinctive modes of vibration that differ from the class of solid or semisolid objects (like doors, wood blocks, and gelatin cubes). An important class of acoustic vibrators happens to include both vibrating strings and vibrating air-filled tubes. A distinctive property of these vibrators is that their partials tend to have frequencies that form a series of simple multiples—that is, their modes of vibration generate a *harmonic series*. Incidentally, the human voice belongs to this class of vibrators.

Most of the common sound sources used to make music produce harmonic partials. Along with the human voice, this includes strings, brass, woodwinds, pianos, organs, guitars, harmonicas, bagpipes, didjeridoos, and many other instruments from around the world. Instruments that produce *inharmonic* partials include drums, timpani, bells, gongs, cymbals, marimbas, glockenspiels, and vibraphones. As a general rule of thumb, if the instrument is played with a mallet or stick, the sound produced probably does not contain harmonic partials (although there are exceptions).

Hearing

Sounds in the world impinge on the eardrum, causing it to vibrate in a way that mimics the vibration of the sound source. The vibrations of the eardrum are then conveyed to a series of three small bones called the *ossicles*, the smallest bones in the human body. The smallest of these, the *stapes*, is just half the size of a grain of rice. The ossicles conduct the sound from the eardrum to the organ of hearing, the *cochlea*.[2] The cochlea is a fluid-filled tube that is curled up so that it resembles a small snail. In fact, *cochlea* is the Latin word for snail. The fluid inside the cochlea moves back and forth in response to the movements of the eardrum.

If we were to uncurl the cochlea, we would end up with a straight tube about 4 centimeters (1.5 inches) long whose diameter is tapered from one end to the other. Sound vibrations enter at the end with the large diameter and travel down the tube toward the smaller end. The small end is called the *apex* and the large end is called the *base*. The tube is divided along its length by a partition that creates two parallel channels. Within this partition is the *basilar membrane*, which contains a dense collection of sensory neurons that connect to the brain via the auditory nerve. The basilar membrane is the most important part of the hearing organ.

Through a tedious series of experiments, the Hungarian physiologist Georg von Békésy discovered that different frequencies stimulate different places along the basilar membrane.[3] Low frequencies are propagated nearly the whole length of the tube and maximally stimulate the most distant points (near the apex of the cochlea). High frequencies, by contrast, do not travel very far and so stimulate points near the base of the cochlea. In effect, there is a mapping between frequency and place of stimulation along the membrane. For this discovery, von Békésy was awarded the Nobel Prize in 1961.

A note played on a piano may generate twenty to thirty partials. These sounds are transmitted through the air to the eardrum, then relayed by the ossicles to the cochlea. Inside the cochlea, the different partials generate different hot spots of stimulation along the basilar membrane. By spreading the frequencies across the membrane, the cochlea attempts to separate out each individual component partial.

If partials are too close together, it will be impossible to detect their separate presence. Two partials need to be at least 1 millimeter apart on the membrane for them to be reliably detected as separate sounds. When the basilar membrane successfully isolates a particular partial, we say that the partial has been *resolved*. Most partials aren't resolved, however. For most sounds, the uppermost partials are too close together for the cochlea to tease them apart. For a typical isolated sound (like a trumpet tone), the basilar membrane may resolve between about five and ten partials. The unresolved partials still contribute to the perception of the sound, but only as a group. For a complicated chord played by an orchestra, hundreds of partials may be present, but the basilar membrane may resolve only twenty-five to thirty of them.

The Packaging of Partials

Oddly, when we listen to a piano tone containing (say) thirty partials, there is nothing in our conscious experience to suggest that there are actually thirty component sounds present. Similarly, there is nothing in our conscious experience to suggest that our cochlea has resolved (say) eight of these partials. We don't hear thirty things or even eight things. Instead, we hear a single thing: a piano tone.

If we open the lid of the piano, we can see two or three strings vibrating for each note. That is, two or three physical sound sources are active when we strike a key on the piano. Still, we don't hear two or three things: we hear one thing, a single sound. The reason is that the brain takes all of the different frequency components and repackages them into something else. Hearing scientists like to call this repackaged thing an *auditory image*. That is, the thirty frequencies generated by three physically vibrating strings (and resolved by the cochlea into perhaps eight partials) ultimately lead to the subjective impression of a single piano tone.

Our experience of the sound is so natural and compelling to us that we mistakenly think we are hearing "the real thing"—what the piano really sounds like. A visiting Martian might find our experience puzzling. The Martian might find it odd that we can't hear that there are actually three piano strings vibrating. Even if we accept the fact that humans can't distinguish the individual vibrating strings, our Martian visitor might still wonder why we can't hear each of the thirty harmonics present in the sound. By talking to a hearing scientist, the Martian might understand that the human basilar membrane is capable of resolving only eight partials for this particular tone. However, if the hearing organ has isolated eight partials in the piano tone, why, the Martian might ask, don't we at least hear the eight resolved harmonics? Eight resolved harmonics are conveyed along the auditory nerve up to the brain, yet our conscious experience is of a single tone rather than eight tones. What is the human brain doing? And why?

There is a good reason for repackaging partials into auditory images. The reason is most obvious when we consider what happens when two unrelated sources generate sound at the same time. Instead of our isolated piano tone, suppose that we heard a piano playing middle C (C4) and a flute playing E5. Together, the two sounds might produce fifty or sixty partials. Perhaps ten or fifteen of these partials will be individually resolved by the listener's basilar membrane. These ten or fifteen partials will be transmitted along the auditory nerve up to the brain. But what is the brain supposed to do with ten or fifteen resolved frequencies?

Brains are practical: they exist to solve practical problems in the world. When interpreting sensations, brains cut to the important stuff: Is that food? Is that my baby crying? Is there any danger here? Where is that sound coming from? What caused that sound?

In the case of sound, it is much more useful for us to perceive "objects" rather than "frequencies." It is much more useful for us to hear "a train whistle" or "a meowing cat" than to identify specific partials. Said another way, the auditory system is designed to recognize sound-producing objects rather than individual modes of vibration. The biological goal here is to create auditory images that represent the actual sound-producing objects in the world. Instead of hearing ten or fifteen individual frequencies, we hear "a piano playing C" and "a flute playing E."

In carrying out this task, the brain has to solve two conceptually separate problems: (1) assembling collections of frequencies into plausible sonic objects and (2) naming or recognizing the physical source for each assembled object. A listener can form an auditory image without necessarily being able to recognize or label the sound. But most of the time we can identify all the sounds we hear.

Brains aren't always successful at this task. That is, they don't always generate auditory images that have a one-to-one relationship with the actual sound sources in the world. For example, twelve violinists playing in unison will not evoke twelve auditory images. Of course, most of us would think that this is not much of a problem. But our Martian visitor might consider this a serious defect in human hearing. Striking an octave on the piano might fool many listeners into thinking that they are hearing a single piano tone. However, many musicians would think that this indicates a deficiency in the listener's skill.

Why do these failures occur? That is, why is the brain not always successful in assembling an accurate picture of the sound-generating objects in the world? Suppose you are the part of the brain responsible for processing sounds. You are looking at ten resolved input frequencies. This might represent ten separate sound sources, each generating a single partial. Or it might be a single sound containing all ten resolved partials. Or it might be five sound-producing objects, each generating two resolved partials. In trying to decipher what might be causing these ten input frequencies, the number of possible combinations turns out to be extraordinarily large.

Hearing scientists refer to the general problem of assembling auditory images from a gaggle of partials as *auditory scene analysis*.[4] In auditory scene analysis, the various partials are grouped together to form plausible auditory images. A common metaphor is to say that the acoustic input is "parsed." In the same way that we parse a sentence into subject, verb, object, and other grammatical elements, the auditory system assigns the partials to different auditory sources. For example, a complex sound scene might be parsed into three images: "slow footsteps over to the right, a telephone ringing behind me, and an aircraft flying overhead." Sometimes identifying labels may not be available to a listener. Where a musician might say, "I hear an oboe, a clarinet, and a bassoon," a nonmusician unfamiliar with instrument names might say, "that nasal sound, that other woody sound, and that third buzzy sound." Both listeners may have successfully parsed the auditory scene, although only the musician has correctly named the sound sources.

The process of auditory scene analysis might be considered successful when each auditory image corresponds to only one actual sound source in the world. An error can occur when several actual sound sources are mistakenly grouped together as a single auditory image or when phantom images are generated that correspond to no actual sound source in the environment.

Notice, by the way, that the biological goal of accurately parsing the auditory scene differs from the musical goal. In music, we may want twelve violins to fuse into a single auditory image, or the pitch sequence sung by a yodeler to break apart and sound like more than one singer.

How does the brain carry out auditory scene analysis? Since the 1970s, researchers have made extraordinary strides in figuring out how the brain does this. The leading researcher in this field has been McGill University psychologist Albert Bregman.[5] Bregman's work suggests that the brain performs scene analysis by relying on a set of heuristics, or rules of thumb. The heuristics are generally successful, but they are also fallible. There is no surefire way to parse an auditory scene into a set of accurate auditory images.

Localization Cues

The flavor of these heuristics can be conveyed by looking at a very simple rule related to *source location*. In general, sounds that come from the same location in space are more likely to come from the same physical sound source. Conversely, sounds that come from different spatial locations are more likely to originate from separate physical sound sources.

Since we are equipped with two ears, the brain has access to some useful localization information. The two principal cues are *time delay* and *intensity*. If a sound source is located to your right, the sound it produces will strike your right ear slightly before your left ear and is likely to be a little louder in your right ear. The brain uses these *interaural* (between-ear) cues to construct a sense of the sound's location.

Localization cues, however, do not provide infallible information for separating sound sources. A simple illustration of how we can be fooled is evident in an ordinary stereo system. When you listen to a stereo, all of the sounds come from the left and right loudspeakers. If you remove the cloth covering the speaker cabinets, you will typically discover that each cabinet contains two physically distinct sound sources: the *woofer* (specialized for low frequencies) and the *tweeter* (specialized for high frequencies). That is, all of the sounds produced by your stereo may be generated by four physical sound sources. Yet when you listen to your stereo, you don't have any sense that there are four physical sound sources.

Stereo technology is intentionally deceptive. The technology has been designed to deceive the human auditory system about the locations of the sounds. However, not all animals will be fooled by a modern stereo system. Bats, for example, have superb localization abilities, so a bat may very well be able to distinguish sounds that originate in the woofer from those generated by the tweeter. Once again, our hypothetical Martian might be surprised that we humans can't tell that all of the sounds are coming from four physical sources. When we listen to a recording, we might hear the singer front and center, the piano left of center, and the acoustic bass right of center. But these are three "virtual" sound sources that don't exist in the real world. In the real world, there are actually four sound sources.

Of course, all this doesn't matter. We don't really want listeners to perceive the real acoustical world of the stereo. We want the stereo's deception to be successful. Ideally, the stereo should act like a sort of transparent sonic window opening onto a fictitious space where the musical sounds are placed. We want people to experience the four parts of a string quartet—first violin, second violin, viola, and cello—not the four parts of the left woofer, right woofer, left tweeter, and right tweeter.

When looking at a mirror, people can often apprehend both the mirror and the virtual images at the same time. That is, a mirror can be perceived as a flat shiny object at the same time that the images "in" the mirror can be seen. In the case of the stereo system, we can sometimes perceive both that sounds are being emitted by a pair of loudspeakers, as well as the virtual sound images that are conveyed. The best sound systems, however, simply evaporate—leaving us with the feeling that the virtual images are the only real ones. Stereo sound is most successful when the loudspeakers themselves disappear from our awareness.

The Pleasures of Stereo

One of the most interesting facts about stereo reproduction is how much nicer it sounds than mono reproduction. When you switch from mono to stereo (or from stereo to mono) there is a direct effect on the experience of pleasure: listeners all over the world prefer stereophonic sound over monophonic sound. Why?

In order to understand why stereo sounds better than mono, we must first consider why pleasure exists. Pleasure—and its opposite, pain—is a powerful motivator. When you accidentally bite your tongue, the experience can be agonizing. But the pain serves a useful purpose: it teaches you to be very careful in coordinating the movements of teeth and tongue. If we didn't experience pain, our tongues would soon be reduced to tatters from constant self-injury.

Like pain, pleasure also serves as an important motivator. Pleasure encourages us to do all sorts of things that are biologically important, like eating, nurturing children, and falling in love. Pleasure and pain provide the carrots and sticks through which biology encourages adaptive behaviors and discourages maladaptive behaviors.

Pain and pleasure are used not simply to punish or reward particular actions but also to punish or reward particular mental behaviors. When you forget a word, you often experience a feeling of mild annoyance or frustration. When you finally retrieve the right word, there is a sense of mild relief or gratification. When you solve a difficult problem, you might have an "aha!" experience that can be deeply satisfying. In each case, the brain provides rewards and punishments to encourage useful mental activity. Without these incentives, our thoughts would simply wander aimlessly.

Apart from encouraging adaptive actions and adaptive thoughts, biological carrots and sticks are also at work in the process of perception. When perceiving the world, the brain is constantly attempting to assemble a coherent scene populated by distinct objects. That is, given all of the sensory elements, the brain is attempting to assemble these elements into a picture where all of the elements are accounted for and all the assembled objects make sense. When this is achieved, we experience a mild sense of pleasure. When we fail at this task, we experience a mild sense of irritation.

When the auditory system is presented with a large number of resolved partials, it faces a considerable challenge in determining how to assign each partial to a particular acoustic source. Localization cues provide one of the helping hands. If frequencies W and X seem to come from the left and frequencies Y and Z seem to come from the right, this is evidence suggesting that W and X may be produced by a single sound source that differs from the sound source producing Y and Z.

When we listen to a mono recording, all of these localization cues are lost. If there is just one loudspeaker, then partials W, X, Y, and Z will all come from the same point in space.[6] The brain must attempt to assign frequency components to different sources without the benefit of localization cues. Does this partial belong to the first violin or to the cello? When the first violin was located off to the right and the cello was located left of center, it was easier for the brain to correctly assign each partial to its true source.

In effect, stereophonic reproduction provides the listener with additional cues that aid in auditory scene analysis. If the brain rewards itself for forming coherent auditory images and if stereophonic localization cues contribute to the listener's ability to form vivid and coherent auditory images, then we can understand why listeners might experience stereophonic sound as sounding more pleasant than monophonic sound.

Now consider the voice-leading rule that forbids parallel octaves. When a solo violin plays an ascending scale, lots of parallel octaves occur as all of the harmonics move upward together in pitch. In fact, research shows that these parallel frequency movements help bind the component partials together so that listeners form a single auditory image of the violin. (We'll discuss this phenomenon more in chapter 6.) But what happens when both the first and second violins play in ascending octaves? At this point, the brain is likely to amalgamate all of the partials from the first and second violins into a single "super-instrument," or *virtual instrument*. By itself, this is not bad. You could have two violins playing a sustained passage in octaves with little perceptual consequence: the brain would simply hear the two instruments as a single auditory image. Like the three strings of the piano tone, most listeners wouldn't be aware of the true number of sound sources. Problems arise, however, when the brain is confused about how to parse the auditory scene. At one moment, the brain may segregate the two instruments as separate auditory images, but at the next moment, the two instruments fuse into a single auditory image. In a passage of otherwise good part-writing, a single parallel octave may sow seeds of doubt in the auditory system about the correct assignment of partials to images. As in the case of a poorly tuned radio that switches from stereo to mono, the momentary parallel octave hampers the efforts of the auditory system to carry out auditory scene analysis. In both cases, there is a loss of pleasure.

The Theater of the Mind

With this background in place, we can now introduce the single most important concept in the field of auditory perception: the distinction between *acoustic* phenomena and *auditory* phenomena. Something is said to be *acoustic* when it pertains to the physical world apart from human beings. For example, when a violin produces a sound, both the vibrating string and the propagated sound are acoustic phenomena. The harmonics or partials that make up the sound are also acoustic phenomena. Something is said to be *auditory* when it pertains to the brain's interpretation of the sonic world. When resolved partials are relayed from the cochlea to the brain, the brain assembles these partials into one or more images, and it is these images that are accessible to consciousness. The violin "out there" is acoustical; the violin in your head is auditory. It should now be clear that you have never heard the true (acoustic) sound of a violin. All you have ever heard are the (auditory) images constructed by your auditory system. In effect, the brain takes the sensory information it receives and creates a sort of theater of the mind where a simulation of the outside world is reenacted on a purely mental stage.[7] If a dolphin or a bat were exposed to the same acoustic violin sound, their brains would create somewhat different theatrical representations of the outside world. The dolphin and bat would have different auditory experiences from us.

Auditory and acoustic phenomena are constantly being confused with each other. In ordinary conversation, the word *sound* is typically used in an ambiguous fashion. We could be speaking of the physical sound generated by a violin or the subjective experience of the auditory image (the "sound") of a violin. When I say, "I hear a violin," I am necessarily speaking of an auditory phenomenon. I might discover that the sound was generated by a computer program simulating the physics of a violin. In this case, I would understand that the sound was not actually produced by a violin. To reduce confusion, it is often helpful to use the modifiers *generated* (for acoustic phenomena) and *evoked* or *perceived* (for auditory phenomena). The violin generates a physical sound. But the physical scraping of a bow on a string evokes the sound of a violin in a person's brain. That is, we perceive a violin. Both the computer program and the actual violin evoke the image of a violin even though the acoustic phenomena are different.

When I say, "I hear Janice playing a violin," what is happening is something along the following lines: Janice is drawing a bow across the string of a physical instrument that people call a violin. The sound is stimulating my auditory system and evoking an image in my brain that I would normally call "the sound of a violin."

This way of speaking might seem ridiculous, but it is musically important. Normally when a (physical) violin is playing, you hear (that is, the sound evokes the experience of) a violin. But there will be times when a (physical) violin is playing yet you will not hear a violin. For example, if a cello and a violin are playing a sustained octave, your brain might misinterpret the partials generated by the violin as belonging to the cello. In other words, your (auditory) experience may be of an especially rich-sounding cello playing a single tone rather than a violin and cello playing two different tones. In one case, the theater of your mind constructs separate images of a cello and a violin. But in the other case, the theater of your mind constructs the image of a sort of cello on steroids. In both cases, however, the acoustic sounds are identical. If we want to understand how people experience sound, it is essential to distinguish acoustic phenomena from auditory phenomena.

Auditory Streams

To this point, I have been speaking as though auditory images are static entities. In vision, objects might move around, but they rarely fade in or out. Sound is different. Unlike physical objects, sounds require a constant infusion of energy in order to maintain their presence. Unlike physical objects, sounds are volatile: they appear and disappear. The sound of dripping water provides a good example of an intermittent sound. Each "drip" sound can evoke an auditory image, but there is also the sense that the drips are connected together in time: they are perceived as a single continuous sound source. As long as the drips aren't separated too much in time, we experience "something dripping," not just the individual drip sounds. In short, sounds don't just form images; they can also form a sense of continuity across time.

Apart from connecting sounds isolated in time, a sense of continuity may also emerge from physically separate sound sources. When you play a one-octave scale on a harp, you are successively activating eight different sound generators, that is, eight independent strings. But the brain doesn't interpret this as eight separate sound sources that happen to follow one after the other. Instead, the brain interprets the sounds as forming a single line.

In 1971, Jeffrey Campbell coined the useful term *auditory stream* to refer to the subjective image of a sound activity continuing over time.[8] An auditory stream emerges when the brain "connects the dots" and forms a continuous image that persists from sound to sound. As in dripping water, that connection may arise from a single physical source that produces intermittent sounds or, as in the harp scale, that connection may arise from interpreting independent sound sources as though they were produced by a single source. In the perception of music, the phenomenon of auditory streams is critically important. Without this psychological phenomenon, there would be no musical lines, just a succession of disconnected sound events with no feeling of connection. Notice that streams are *auditory* phenomena: the sense of connectedness may be entirely an auditory illusion with no acoustic reality.

As we will see in later chapters, stringent conditions determine whether we hear discrete sounds as connected. That is, we will discover that lines of sound (such as melodies) form only when certain criteria are met. We will also discover that when several concurrent lines of sound are present (as in part-writing), the potential for auditory confusion increases dramatically. We will see that the practice of voice leading provides a sonic toolkit that helps us keep control over auditory confusion.

The Pleasures of Puzzle Solving

Throughout the remainder of this book, I will often compare auditory scene analysis to working on a jigsaw puzzle. The puzzle pieces are the resolved partials that must be assembled into a coherent picture. Ideally, this process will result in a set of auditory images where each image corresponds to an actual acoustic source in the environment. So how does the brain know when it has successfully solved the puzzle? One clue is that there are no left-over puzzle pieces—there are no orphan partials that haven't been assigned to an auditory image. Another clue is that all the puzzle pieces fit snugly with no forced connections: that is, each partial combines easily with the other partials that make up a given auditory image. Yet another clue is that there is no indecision about whether a particular puzzle piece is in its correct location: that is, there is no flip-flopping about assigning a partial first to one image and then to another. A final clue is that the overall picture makes sense. For example, images that existed a moment ago don't inexplicably disappear. If a person is able to *see* the sound sources, then the visual and auditory perceptions can be compared: the auditory scene analysis might be deemed a success if the number and apparent locations of the auditory images do not contradict what a person is seeing.

Reprise

When someone says, "I hear the sound of a squeaking chair," this simple statement masks a remarkably sophisticated mental achievement of packaging and labeling. In this chapter, we have noted that acoustic sources typically vibrate in several *modes* concurrently. Each mode produces a distinctive component sound, called a *partial*—with its own distinctive *frequency*. When a sound source produces two or more partials, the result is referred to as a *complex tone*. The sound energy is communicated through the air to the eardrums, and the flapping eardrums in turn relay this energy through the *ossicles* to the *cochlea*. Inside the cochlea, different frequencies cause different points of maximum stimulation along the *basilar membrane*. If two frequencies activate points that are sufficiently far apart, the corresponding partials will be resolved as separate signals. *Resolved partials* are transmitted up the *auditory nerve* to the brain, where they provide the main puzzle pieces used to assemble *auditory images*.[9] Ideally, this process will result in a set of auditory images where each image corresponds to an actual acoustic source in the environment. Acoustic sources (that exist in the physical world) are translated into auditory images (that exist only in the mental world); *acoustic scenes* (potentially containing several physical sources) are reformulated in the theater of the mind as *auditory scenes*. When an auditory image is maintained over the course of time, the sound events are heard as connected, forming a sense of sonic line or *auditory stream*. Successful *scene analysis* is thought to evoke a sense of pleasure: a mental reward for using all of the puzzle pieces and assembling coherent auditory objects. The success (or lack of success) of this process can lead to positive (or negative) feelings. We will return to discuss the aesthetics of image formation in chapter 16.

4 Principles of Image Formation

In the previous chapter, we introduced the concept of an auditory image and the process of auditory scene analysis. Experimental research has identified a number of factors that contribute to this process. In this and the next two chapters, we will identify several perceptual principles that influence image formation and discuss how these principles relate to music.

Harmonicity

Acousticians distinguish three general classes of sounds: harmonic, inharmonic, and aperiodic. As we learned in the last chapter, *harmonic* sounds contain partials that are related by simple integer frequency ratios. For example, the frequencies 300 Hz, 600 Hz, 900 Hz, 1,200 Hz, 1,500 Hz, ... form a harmonic series. Examples of harmonic sound sources include trombones, fiddles, lutes, bass guitars, pan pipes, and mandolins. *Inharmonic* sounds are sounds whose partials are not related by simple integer frequency ratios. For example, the frequencies 100 Hz, 113 Hz, 240 Hz, 385 Hz, 429 Hz, ... form an inharmonic set of partials. Examples of inharmonic sound sources include wood blocks, tambourines, xylophones, bass drums, steel pans, and bells. Finally, *aperiodic* sounds contain no stable frequencies; their frequency content is chaotic and commonly consists of bands of noise. Examples of aperiodic sound sources include waterfalls, radio static, air conditioners, hissing snakes, whispered speech, and cymbals.

Suppose your auditory system received the following set of resolved partials from the cochlear nerve:

120 Hz   240 Hz   360 Hz   480 Hz   600 Hz   720 Hz   840 Hz

Clearly all seven resolved partials are harmonically related: they are all multiples of 120 Hz, the fundamental for this set. In packaging these partials, it makes sense to assume that they are all harmonics from a single sound source in the environment. Consequently, the auditory cortex ties a metaphorical string around these partials and presents them to conscious awareness as a single sound.

Consider now some more complicated cases. What do you suppose the auditory system would make of the following resolved partials?

400 Hz   600 Hz   1,000 Hz   1,200 Hz   1,600 Hz   1,800 Hz

This looks like a single harmonic series except that some of the harmonics are missing. Specifically, all of the partials are multiples of 200 Hz, but 200, 800, and 1,400 are absent. These partials might be missing in the original sound, or they may have failed to be resolved by the basilar membrane. Although the picture seems incomplete, once again it looks as if the most sensible solution would to be to regard all of the partials as harmonics of a single sound source. In this case, the brain would indeed construct a single auditory image for consciousness.

Now consider the resolved partials:

150 Hz   205 Hz   300 Hz   410 Hz   450 Hz   600 Hz   615 Hz

This looks like two harmonic series. One has a fundamental of 150 Hz and includes the resolved partials of 300 Hz, 450 Hz, and 600 Hz. The other series has a fundamental of 205 Hz and includes the harmonics 410 Hz and 615 Hz. For this input, a reasonable solution would be to present two images to consciousness: one consisting of four partials and a second consisting of three partials.

In the following set of resolved partials, there is an obvious harmonic series involving the frequencies 110, 220, 330, 440, and 550:

110 Hz   220 Hz   258 Hz   330 Hz   440 Hz   455 Hz   487 Hz   593 Hz

The frequencies 258, 455, 487, and 593 are a mystery: they are not multiples or even near multiples of each other. It could be that all four of these misfit partials are components of a single inharmonic sound source. Or perhaps there are two inharmonic sounds, each containing two resolved partials. Another possibility is that there are four unrelated pure tones present in the environment. (Since lone pure tones are rare in nature, this is much less likely.) The auditory system must do something with this acoustic scene.

One solution would be to present two images to consciousness: the first is a harmonic sound source with a fundamental of 110 Hz (the four partials would form a nice clear image); the other four partials are assembled into a second image (of a single proposed inharmonic sound). Notice, however, that we are far more confident of the image with a harmonic series than we are of the proposed inharmonic set of partials. Our metaphorical string tied around the inharmonic partials is rather lose and threadbare. When passing this assemblage up to consciousness, it would be nice if we could also convey our lack of confidence. As we'll see later in this chapter, there is indeed a signal that effectively says to the auditory system, "We're not so sure about this image."

A final complication is that harmonic partials are often slightly out of tune. In the following case, the harmonics are flattened by about 1 percent:

100 Hz   199 Hz   298 Hz   397 Hz   496 Hz   595 Hz

It is also common to see slightly sharp harmonics. For example, most piano strings produce moderately stretched harmonics. When harmonics are mistuned by more than about 2 or 3 percent, they are more likely to stick out and be perceived as separate sounds.

In all of these examples, the harmonic series provides a useful tool for helping to package resolved partials into images. The idea that the auditory system employs a harmonic template in order to assemble images is referred to as the *harmonic sieve*. The technical basis for this idea was worked out in the 1980s by the Dutch scientist Hendrikus Duifhuis and refined by his student Michaël Scheffers.[1] In the late 1990s, two British researchers, Jeffrey Brunstrom and Brian Roberts, published a set of experiments that hearing scientists regard as definitive support for the harmonic sieve concept.[2]

As we've seen, harmonic sounds are not always well behaved. Individual harmonics may be missing, or they may be mistuned by greater or lesser amounts. The degree to which a set of partials conforms to an ideal harmonic series is referred to as its *harmonicity*. A perfectly tuned harmonic series has a high harmonicity; a series containing mistuned or missing harmonics has a lower harmonicity; a series of inharmonic partials has an even lower harmonicity; and noise bands have the lowest harmonicity. Note that harmonicity is an *acoustical* concept used to characterize the physical relationship between a set of partials. There is a parallel auditory concept to harmonicity, *harmonic fusion*.

Harmonic Fusion

Suppose we generate two sounds separated by an exact octave. The sounds might have fundamental frequencies of 111 Hz and 222 Hz, respectively. What happens when we amalgamate the first five harmonics for both sounds?

Tone 1 (Hz)	111	222	333	444	555			
Tone 2 (Hz)		222		444		666	888	1,110
Combination	111	222,222	333	444,444	555	666	888	1,110

Two of the partials are duplicated (222 Hz and 444 Hz). The corresponding points are stimulated along the basilar membrane, but the membrane has no way to distinguish whether a point was activated by one sound source or by several sound sources producing the same frequency. As a result, the auditory nerve conveys just eight resolved partials, from 111 Hz to 1,110 Hz. To the brain, this looks like a single complex tone with a fundamental of 111 Hz. Although there are actually two acoustical sound sources, the brain constructs a single auditory image.

Notice, incidentally, that a little mistuning might help to segregate these two sound sources into separate auditory images. As noted above, a 2 or 3 percent change in tuning will help make a harmonic stick out. What we mean by "stick out" is that the brain forms a distinct auditory image that consciousness can then attend to.

Another pertinent point is that in the real world, vibrating objects rarely have stable frequencies. Even for a seemingly steady instrument tone, the individual frequencies tend to wobble up and down to some extent. As we will see later, these changes in frequency can also affect how the brain assembles partials into auditory images.

The example above demonstrates that it is possible for two independent acoustical sound sources to be interpreted as a single auditory image. We call this phenomenon *harmonic fusion*.[3] Notice that harmonic fusion is a type of perceptual mistake. If the biological goal of the auditory system is to represent each acoustic source by its own auditory image, then harmonic fusion means the system has failed: the brain has mistakenly combined two things that in reality are separate. Harmonic fusion arises because the auditory system relies on the harmonic sieve when assembling auditory images. If the partials form a nice harmonic series (even if they come from different acoustic sources), they are likely to be grouped together into a single image. In short, the harmonic sieve provides a useful—but fallible—tool for assembling auditory images.

When two harmonic complex tones are present, the likelihood of their fusing depends on the frequency ratio separating the two fundamentals. That is, the probability of harmonic fusion depends primarily on the interval between the fundamentals. The interval that most encourages harmonic fusion is the aptly named *unison* (literally, "one sound"). Obviously if both sources produce a harmonic series, all of the harmonics will coincide in frequency. The multiple strings for each note on the piano are tuned in unison, so when we strike a single key, we experience a single auditory image rather than hearing two or three distinct vibrating strings.

The second most fused interval is the octave. The harmonics of the upper note of the octave correspond to all of the even-numbered harmonics of the lower note. This effect is commonplace in pipe organs. If we select an 8-foot and 4-foot stop, each key press will result in two pipes sounding simultaneously—an octave apart. Yet listeners typically experience a single auditory image rather than hearing the two pipes separately. At times, it is possible for listeners to "hear-out" the two pipes. This is called "analytic listening" and is discussed in chapter 13.

The third most fused interval is the fifteenth, or double octave. Here the fundamental of the upper note is the same as the fourth harmonic of the lower note. The second harmonic of the upper note aligns with the eighth harmonic of the lower note, and so on. Harmonic fusion is greater if the partials of the upper note are generally quieter than the partials of the lower note.

The next most fused interval is the perfect twelfth (an octave plus a perfect fifth). In the following example, the harmonics of the upper note align with every third harmonic of the lower note.

Tone 1 (Hz)			300			600			900
Tone 2 (Hz)	100	200	300	400	500	600	700	800	900

After the twelfth, the next most fused interval is the perfect fifth. In this case, the even harmonics of the upper note align with the odd harmonics of the lower note (see below). However, the odd harmonics of the upper note will not coincide with any harmonics of the lower note. For example, the third harmonic of the upper note and the fourth harmonic of the lower note will be offset by two semitones. Nevertheless, the two sets of partials can be interpreted as harmonics of a single complex tone whose fundamental is missing. (In the example below, the missing fundamental would be 50 Hz.) In these cases, the amount of harmonic fusion will depend primarily on the timbre of the participating tones and how high the two tones are in overall pitch. Harmonic fusion is greater when there is less energy present in the partials that don't conform to the harmonic series of the lower tone and if the overall tessitura of the tones is relatively high. Compared with the unison, octave, and twelfth, the perfect fifth is less likely to cause harmonic fusion, and if harmonic fusion results, the harmonicity of the aggregate sound is lower.

Tone 1 (Hz)		150		300		450		600		750
Tone 2 (Hz)	100		200	300	400		500	600	700	

In general, harmonic fusion occurs when the combined partials from two (or more) acoustic sources are aligned like the partials of a single hypothetical harmonic series. This is most likely to occur when the fundamental frequencies of the component tones are related by simple integer ratios.[4] But the degree of harmonic fusion is also affected by the harmonic content (timbre) of the participating tones.

In light of the research on harmonic fusion, we can formulate the following principle:

1. Harmonic Fusion Principle

Partials are more likely to be assembled into a single auditory image when they conform to a harmonic series. If two complex tones are related by simple integer frequency ratios, their partials are likely to align and so fuse as a single auditory image. Intervals that promote such fusion include (in decreasing order) unisons, octaves, fifteenths, perfect twelfths, and perfect fifths. A composer who aims to ensure the perceptual independence of concurrent sounds should shun harmonic intervals in direct proportion to the degree to which they promote the formation of a single auditory image.

We'll see many musically pertinent examples of this principle in the ensuing chapters. For now, let's continue with a discussion of our second principle.

Toneness

When the auditory system assembles a good auditory image, we often experience it as a *tone*. Not all auditory images sound like tones; for example, the sound of flowing water is not very tonelike. However, a set of harmonic partials (like the harmonics produced by an oboe) typically sounds like a "tone."

Sounds can be more or less tonelike. If you know how to whistle, you can illustrate this for yourself. Begin by whistling the clearest tone you can. You'll find that your lips form a small opening that is almost perfectly circular. Now open your lips very slightly. You'll hear that your tone sounds just a wee bit breathy. Now open your lips a little more. As you continue making sounds, you'll hear that as your mouth gets wider, the pitch gets fuzzier and fuzzier. Finally, with your mouth wide open, you can still make a sound that has only the slightest hint of a pitch remaining; mostly you hear noise. Each successive sound is less and less tonelike. At the same time, the pitch of the sound becomes progressively less well defined with each successive sound. What starts off as a clean or clear sound becomes fuzzy—as though the sound were out of focus.

Notice that there seems to be a link between three properties. When a sound is *clear*, it also sounds more *tone*like; at the same time, it also produces a better-sounding *pitch*. Conversely, fuzzy sounds are less tonelike and produce less defined pitches. Psychoacousticians use the word *tonality* to refer to how "tonelike" a sound is. The sound of flowing water has low tonality, whereas a bowed cello note has high tonality.[5] For musicians, the choice of the term *tonality* is unfortunate. (We prefer to use the word *tonality* for other purposes.) So music researchers prefer to use the word *toneness* to refer to how tonelike a sound is.

In the case of whistling, we can reduce the toneness by adding noise (making the whistle breathier). As we add more noise, we transform a harmonic sound into an aperiodic sound. Recall that there is a third type of sound: the inharmonic sound. A classic example of an inharmonic sound is the sound of a bell. Unlike a tone played on a piano, the sound of a bell produces more than one noticeable pitch. If you listen closely, you can hear some of the different pitches. In fact, bell makers have names for the different pitches: the three most prominent pitches are the *hum tone*, the *ring tone*, and the *strike tone*. Anything that sounds clangerous usually evokes more than one pitch. If a sound source evokes more than one noticeable pitch, the sound is typically inharmonic.

By way of summary, I've discussed three different kinds of sounds: harmonic, inharmonic, and aperiodic. Harmonic sounds typically generate clear pitches and have high toneness. Aperiodic sounds typically generate little sense of pitch and have low toneness. Inharmonic sounds typically generate more than one noticeable pitch, leading to a sense of "competing pitches" or "ambiguous pitch"; they have an intermediate toneness.

It turns out that toneness is also influenced by register (high or low). Frequencies above about 5,000 Hz tend to sound like indistinct "sizzles."[6] While most people can easily hear a frequency of 10,000 Hz, the resulting sound doesn't produce a very good sense of tone and also doesn't produce much of a sense of pitch. Similarly, very low frequencies sound like vague rumblings: there's not much toneness, and the pitch is typically obscure.

As a simple demonstration, we can play pure tones of different frequencies to listeners and ask them a sort of Goldilocks question: Is this sound too high? Is this sound too low? Is this sound just right? When asked to judge pure tones, people tend to cluster their "just right" answers around F5 (about 700 Hz)—the F at the top of the treble staff. When we repeat this task with complex tones (containing several harmonics), people tend to cluster their "just right" answers around D4 (about 300 Hz)—the D just above middle C. Since music is created using complex tones, the musically pertinent answer leads us to D4. When listening to complex tones, people find the pitches around D4 to be "just right."

Before continuing, let's pause and take stock. We have discussed three properties—*clarity*, *toneness*, and *pitch*—that are very closely linked. A clear sound often has high toneness and evokes a good sense of pitch. We've also learned that these properties are related to the type of sound: harmonic sounds have the highest toneness, inharmonic sounds have lower toneness, and aperiodic sounds typically have the lowest toneness. Finally, we've learned that toneness is not very good in the high- and low-frequency regions. There is a central region in which toneness (and clarity and pitch) is best.

More formal versions of the Goldilocks experiment have been carried out, and psychoacousticians have created sophisticated models of how the auditory system processes pitch. A classic model was devised by Ernst Terhardt and his colleagues at the Technical University of Munich.[7] Given any

Principles of Image Formation

sound input, Terhardt's model predicts which partials will be assembled together and which tones are likely to be perceived by a listener. It also predicts the clarity of each perceived tone. In short, Terhardt's model can be used to identify the likely toneness for each auditory image. Using the model, we will find that a trumpet has greater toneness than a bell and a bell has greater toneness than the sound of a refrigerator.[8]

As already noted, toneness also changes with pitch register. This is apparent in figure 4.1 where changes of toneness are plotted against pitch for several hundred complex tones.[9] The toneness values were calculated using Terhardt's model. The solid curve shows the toneness for standard harmonic tones ranging from C1 to C7, where C4 is middle C. The dotted curves show changes of toneness for recorded tones spanning the entire ranges for six orchestral instruments: harp, violin, flute, trumpet, cello, and contrabassoon. Although the timbre of the individual notes influences toneness, figure 4.1 shows that the region of maximum toneness for complex tones remains quite stable no matter what instrument is playing. Maximum toneness spans a broad region centered around D4.

Notice that extremely high and extremely low pitches show pretty low toneness. Suppose we wanted to avoid using these tones when making music. For example, we might choose to use only tones that score 1.0 or greater on the scale produced by Terhardt's model. Looking at the graph, we can see that such tones would fall between about E2 and G5, a range that coincides almost exactly with the range spanned by the bass and treble staves in Western music.

Richard Parncutt, at the University of Graz in Austria, and I calculated the average notated pitch in a very large sample of notes drawn from various musical works: Western instrumental music, as well as Korean, Chinese, and Japanese instrumental works. We intentionally excluded vocal music. We found that the average musical pitch for this cross-cultural sample lies near D#4, just a semitone away from the pitch (D4) that experimenters have found evokes maximum toneness.[10]

This suggests that musicians have been attracted to use the tones that evoke the best pitches—and as we've seen, the best pitches correspond to the clearest auditory images. All frequencies are created equal, but not all pitches are the same: some pitches are better than others, and the really good ones are evoked by harmonic complex tones centered near D4. Middle C really is in the middle of something. It's in the middle of the region

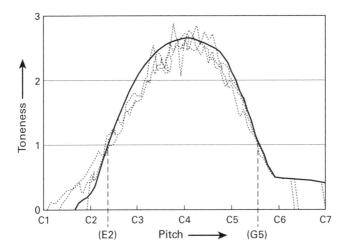

Figure 4.1
Changes of toneness versus pitch for complex tones from various natural and artificial sources, calculated according to the method described in Terhardt, Stoll, and Seewann (1982a, 1982b). Solid line: pitch weight for standardized electronic tones from C1 to C7. Dotted lines: toneness for recorded tones spanning the entire ranges for harp, violin, flute, trumpet, cello, and contrabassoon.

of good toneness; it's in the region where human brains form the clearest auditory images.

(By the way, middle C is *not* in the middle of human hearing. The range of human hearing is usually given as 20 Hz to 20,000 Hz, but these numbers are crude approximations. Young children are regularly able to hear 30,000 Hz and higher. Our ability to hear high frequencies declines rapidly with age, so by the age of twenty-five, few people can hear much beyond 16,000 Hz. Similarly, few people can hear as low as 20 Hz unless it is very loud. In frequency terms, the "center" of hearing for most people would be around 8,000 Hz. In logarithmic [or octave] terms, the "center" of hearing would be around 1,000 Hz. Middle C has a frequency of about 262 Hz, which means it is very low compared with the range of human hearing. Middle C is nowhere near the middle of hearing: instead, it's near the middle of the range for good pitch perception.)

In all of this discussion, notice the special status of harmonic sounds. Harmonic sounds tend to produce partials that are (1) easier for the auditory system to assemble into auditory images, (2) produce a subjectively clearer or cleaner sound, (3) sound more tonelike, and (4) produce clearer, less ambiguous pitches. Why does this matter? If you play three notes simultaneously on a piano, it is relatively easy to hear-out the three distinct pitches, especially if the notes are widely spaced. Now imagine three different cars parked beside each other, each with its engine running. Here it is much more difficult to hear-out the sounds of the individual cars. The engine sounds are more amorphous than the tones of a piano. Similarly, if you simultaneously turn on an electric fan, an electric mixer, and a dishwasher, it is hard to distinguish each individual sound source. By contrast, simultaneous sounds produced by a bassoon, French horn, and clarinet are relatively easy for the auditory system to discriminate from each other. In short, harmonic tones are easier to segregate perceptually compared with inharmonic or aperiodic sounds.

It bears emphasizing that there is absolutely nothing wrong with using inharmonic or aperiodic sounds in music-making. Musicians are entirely free to use the sound of wind blowing through trees (an aperiodic sound) or banging pots and pans (inharmonic sounds). But one of the repercussions of using such sounds is that it will be harder for listeners to maintain perceptually independent auditory images for these sounds *when such sounds are generated concurrently.*

As we've just seen, past music-making has favored the use of harmonic tones and has favored tones in the region of greatest pitch clarity. Said another way, over the centuries and across nearly all cultures, musicians have tended to be attracted to those sounds that happen to evoke the best auditory images.

Drawing on the existing research concerning pitch perception, we can formulate the following principle:

2. Toneness Principle
Clear auditory images are evoked when tones exhibit a high degree of toneness. High toneness is associated with tones that evoke clear pitches. Tones having the highest toneness are harmonic complex tones centered near D4 (approximately 300 Hz) in the region between E2 (bottom of the bass staff) and G5 (top of the treble staff). Tones having inharmonic partials produce competing pitch perceptions, and so evoke more ambiguous auditory images. The weakest auditory images are produced by aperiodic noises.

At the beginning of this chapter, we saw that it's not always clear how partials should be grouped into auditory images. When the partials conform to a harmonic series, the auditory system can be pretty sure that they belong together. But as the partials show less and less harmonicity, the auditory system is less and less confident about how they should be grouped. We also noted that it would be useful if the auditory system could signal a degree of confidence when passing images up to conscious awareness ("I'm pretty sure about this image; I'm not so sure about this other one"). We've seen that sound images differ in clarity, toneness, and pitch. It makes perfect sense, then, that the sounds that are easiest to assemble into auditory images are those that listeners would describe as sounding "clearest." These same sounds are also the most tonelike—that is, they exhibit the greatest toneness. In short, clarity and toneness fit the bill as possible signals of confidence in auditory scene analysis.

So what about pitch? Is pitch just another signal of confidence like clarity and toneness? Unlike clarity and toneness, pitch has a *domain*; it can span a wide range of values, which in many Western languages is described using a height metaphor of *low* to *high*.[11] We hear this property apart from whether the pitches sound "better" or "worse." So why do we experience a sense of pitch?

We musicians tend to think of pitch as an objective property. You strike a particular key on the piano, and it generates a particular pitch. Pitch looks like a property of the sound itself—an *acoustical* property. Alas, this turns out to be wrong. We don't have the space here to review the evidence, but hearing scientists have established quite convincingly that pitch is an auditory phenomenon, not an acoustic phenomenon. Despite appearances, pitch doesn't exist in the external world; it isn't "out there." Instead, pitch is constructed in listeners' heads. Pressing a piano key generates a bunch of partials, and the auditory system uses these partials to construct the experience of a particular pitch. Said another way, instruments don't *generate* pitches; they *evoke* pitches in brains. When I first encountered this idea, I thought that it was patent nonsense, so I understand why some readers might be skeptical. The subjective origin of pitch is best demonstrated through audio examples, so skeptical readers are encouraged to refer to the Web for further clarification.

Since pitch is a creation of minds (rather than facts about the external world), we might ask why the brain creates pitch. What is the biological purpose of pitch? In chapter 3 we learned that the auditory system engages in two important tasks: assembling partials into images and then labeling or identifying each presumed sound-producing object. We can name most of the sounds we hear: that's a bird calling, that's flowing water, that's a clarinet tone, and so on. But notice that all of these examples rely on language. What about sounds for which we don't have words? (A trivial example is a sound you've never heard before.) When a set of partials forms a nice harmonic series, the brain can be pretty confident that all the partials belong together in a single image, even if the conscious brain has no idea what to call it. *Pitch* appears to be the brain's prelinguistic default label. We do not need language in order for consciousness to identify or tag different sound sources.

Notice, however, that these pitch labels are evoked only for images with high toneness—those images about which the auditory system is most confident. When assigning pitches to images, it is as though the auditory system follows the principle: "Don't bother labeling a sound unless you're convinced it's real." More technically, "Evoke pitch only for those auditory images that have a high probability of corresponding to a real acoustic source."[12]

Reprise

As listeners, we are almost never aware of the existence of individual partials. Instead, we have conscious access only to the auditory images that the brain has assembled from the various partials resolved on the basilar membrane.[13] So how does the brain assemble images from resolved partials? In this chapter we have seen that it is easiest to form images when partials conform to the harmonic series. That is, images are most easily formed when the partials exhibit high *harmonicity*. In assembling these images, the auditory system employs a *harmonic sieve*—a template that is forgiving of missing harmonics and also tolerates a small degree of mistuning. Although *harmonicity* offers a useful rule of thumb for grouping partials, it can also lead to *harmonic fusion* when two or more independent sound sources happen to produce harmonically related tones.

In the case of *inharmonic* and *aperiodic* sounds, assembling auditory images is much more difficult. It is helpful if the auditory system signals the degree of confidence for each image. Subjectively, poor auditory images are experienced as less tonelike, or less "clear" sounding.

The clearest auditory images evoke a sense of *pitch*. The phenomenon of pitch appears to be the brain's default prelinguistic way of labeling a sound. When there are several concurrent sound sources, it is easier to hear-out images that exhibit clear pitches. Said another way, we can most easily decipher an acoustic scene when the constituent sources produce harmonic partials. Even for harmonic sound sources, some sounds sound clearer than others. As we've seen, the clearest tones evoke pitches around D4—in the region near middle C.

5 Auditory Masking

Standing in the shower, you pause momentarily and listen: *Is that my telephone ringing?* In noisy environments, otherwise easily heard sounds can become difficult or impossible to detect. This interference is called *masking*. Masking is an auditory phenomenon, not an acoustic phenomenon; in the real world, sounds rarely obscure each other.[1] All of the masking occurs in your head. Our hypothetical Martian visitor might have no difficulty hearing your telephone ringing, no matter how loud the surrounding sounds are. But that's not the way human hearing works.

Masking is most noticeable when the interfering sounds are loud, but in fact, masking occurs all of the time; we just aren't always aware of its effects. Even the quiet hum of a refrigerator produces masking, but the amount of masking may not be sufficient to entirely drown out other sounds that may be present. We call this *partial masking*. When a full orchestra is playing, nearly all of the sounds produced by the various instruments (thousands of partials) are fully masked. What we hear is a small subset of partially masked sounds that manage to escape from the cacophony of acoustic activity. Hearing scientists have succeeded in identifying the physiological basis for masking. Understanding some of the pertinent physiology will prove surprisingly helpful in explaining several important musical phenomena. We'll also see that the physiology of masking has had important social repercussions—notably, it has had a dramatic impact on the participation of women in music-making.

In chapter 3, I noted that different frequencies produce different points of maximum stimulation along the basilar membrane in the cochlea. Low frequencies cause the greatest stimulation of the membrane near the far end (apex), whereas high frequencies produce maximum stimulation near the cochlea's entry point (base).[2] The entire basilar membrane is roughly

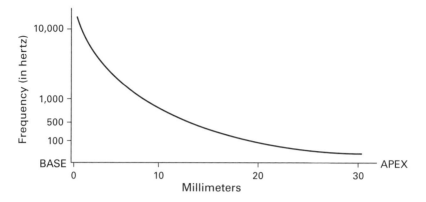

Figure 5.1
Relationship between input frequency and place of maximum stimulation on the basilar membrane.

30 millimeters in length.³ This means that the entire range of humanly detectable frequencies is mapped onto a patch of tissue whose total length is about the width of a ruler. Figure 5.1 plots the relationship between place of maximum stimulation and input frequency.

When concurrent pure tones have the same frequency, they stimulate the same point on the basilar membrane and so are resolved as though only a single partial is present. When two concurrent frequencies are close but not identical, they stimulate neighboring regions along the basilar membrane. When the stimulation points are close enough, the two frequencies generate mechanical interference that results in masking.

Working at the Bell Telephone Laboratories in the 1950s, Harvey Fletcher carried out experiments to measure how much frequency separation was required between two (pure) tones in order to avoid masking. He manipulated the frequencies of two tones until the point where they just began to cause masking. This minimum frequency separation Fletcher called a *critical band*. When two pure tones lie within a critical band of each other, they produce a measurable amount of mutual masking. When separated by more than a critical band, pure tones do not interfere with each other.⁴

Fletcher discovered that the size of critical bands is not constant over the range of hearing. The change in size is illustrated in figure 5.2, which plots a succession of tones, each separated by a critical band from its neighbors. Around middle C, the critical band is roughly three semitones (a minor third) in size. As the tessitura gets higher, critical bands become slightly

Figure 5.2
Approximate size of critical bands represented using musical notation. Successive notes are separated by approximately one critical bandwidth—roughly 1 millimeter separation along the basilar membrane. Notated pitches represent pure tones rather than complex tones. *Note*: The internote distances plotted in this figure have been calculated according to the equivalent rectangular bandwidth-rate (ERB) scale devised by Moore and Glasberg (1983; revised Glasberg & Moore, 1990).

smaller—to around two semitones in size. As the tessitura descends below middle C, the size of critical bands increases quite quickly.

It is important to understand that the tones notated in figure 5.2 represent pure rather than complex tones. Since pure tones are virtually never used in music, musicians can easily get the wrong impression when looking at this figure. Any normal tone will contain many partials, and the partials will be distributed over many critical bands.

Figure 5.3 shows the harmonics of a complex tone with a pitch of C2 (at the bottom of the bass staff). The first harmonic (C2) has a frequency of about 65.4 Hz. Successive harmonics have frequencies of 130.8 Hz (C3), 196.2 Hz (near G3), 261.6 Hz (C4), 327.0 Hz (near E4), 392.5 (near G4), and so on. Notice that the seventh and eighth harmonics (near B♭4 and C5) are separated by only about two semitones. However, from figure 5.2 we can see that a critical band in this region is about three semitones in width. This means that the seventh and eighth harmonics are close enough to interfere with each other on the basilar membrane. Although these two harmonics belong to the same complex tone, they will engage in mutual masking. If one of these harmonics is significantly louder than the other, it might completely mask the presence of the other harmonic. If the harmonics have similar amounts of energy, then neither will be resolved clearly. As we continue up the harmonic series, the successive harmonics get closer and closer, increasing the likelihood that two or more harmonics lie within a critical band of each other, and so become difficult to resolve on the basilar membrane. For the complex tone shown in figure 5.3, the likelihood is that only six partials will be resolved: harmonics 1 through 6.

Figure 5.3
The approximate position of the first sixteen harmonics for a single complex tone with a fundamental of C2 (roughly 65 Hz). Compare the distances between successive harmonics with the size of critical bands shown in figure 5.2. Successive lower harmonics are separated by more than a critical band and so will be individually resolved by the cochlea. Beyond the sixth harmonic, the distances separating successive partials are smaller than a critical band, so neighboring harmonics will tend to interfere with each other. Notated pitches represent pure tones rather than complex tones.

Working at Harvard University in the 1960s, Donald Greenwood compared measures of critical bandwidth with distance measurements along the basilar membrane.[5] He showed a simple linear relationship, with one critical bandwidth being roughly equivalent to the distance of 1 millimeter along the membrane. In other words, pure tones tend to begin masking each other when their points of stimulation on the basilar membrane are less than 1 millimeter apart. This means that each successive (pure) tone notated in figure 5.2 is separated by about 1 millimeter along the basilar membrane. You might think of the tones notated in figure 5.2 as fence posts marking out equal distances along the membrane. Although the pitch intervals vary in size, the successive fence posts are all the same physical distance apart. For example, the distance between C2 and G#2 is the same as the distance between D4 and F4—each interval represents about 1 millimeter along the membrane. Once again, readers need to remember that the notated tones in figure 5.2 represent single pure tones. Usually a notated note represents the fundamental of a complex tone containing many harmonics.[6]

In 1965, Dutch researchers Reinier Plomp and Willem Levelt suggested that musicians tend to spread the sound energy evenly across the basilar membrane. A useful analogy here is to think of the basilar membrane like a piece of toast. When spreading jam on toast, some people slap on the jam in indiscriminate clumps. Others like to spread their jam carefully so it covers the entire surface of the toast in a uniform layer. If partials clump together on the basilar membrane, then the clumps will generate lots of mutual interference. So if we want to reduce auditory masking, it helps to spread the sound energy evenly across the membrane. In figure 5.2, this means that a musician should aim to have a similar number of partials in each of the successive intervals. That is, there should be roughly the same number of partials between C2 and G#2 as between G#2 and D3, and so on all across the range of hearing.

In order to test this idea, Plomp and Levelt carried out a study of two musical works: the third movement from J. S. Bach's Trio Sonata No. 2 for organ and the third movement from Dvořák's String Quartet in E♭ major, op. 51. First, they rewrote the chords so that the first nine harmonics were included for all tones in the chord. In this way, something notated as a chord consisting of four (complex) notes would be rewritten as a sonority consisting of 36 (pure-tone) partials. Then they measured the spacing between all pairs of successive partials and compared these distances to the size of the critical band in that range of hearing. Their analysis showed that both Bach and Dvořák arranged the vertical sonorities in a way consistent with an even spacing across the basilar membrane: roughly the same number of partials occur in each critical band. In 1992, Peter Sellmer and I published a study that tested Plomp and Levelt's idea more thoroughly using a more refined method and a much larger musical sample. Our results provided even stronger evidence that composers arrange chords in a way that tends to spread the partials evenly across the basilar membrane. Like a finicky person spreading jam on toast, composers tend to space out the tones in chords so that the sound energy of the partials is more or less evenly spread across the organ of hearing.

This effect is difficult to illustrate graphically. However, figure 5.4 provides a useful illustration. The figure shows the average spacing of notated (complex) tones for sonorities having various bass pitches from C4 down to C2. For example, the first notated sonority in figure 5.4 shows the average tenor (E4), average alto (A4), and average soprano (D#5) pitches for a large sample of four-note sonorities having C4 as the bass pitch. (The specific chords notated in figure 5.4 should not be interpreted literally; no composer wrote these specific chords. Only the approximate spacing of the voices is of interest.) As the tessitura of the chords descend, the pitch separation between the lower notes in the sonority tends to become larger. This

Figure 5.4
Average spacing of tones for sonorities having various bass pitches from C4 to C2. Calculated from over ten thousand four-note sonorities extracted from Haydn string quartets and Bach keyboard works. Bass pitches are fixed. For each bass pitch, the average tenor, alto, and soprano pitches are plotted to the nearest semitone. (Readers should not be distracted by the specific sonorities notated; only the approximate spacing of voices is of interest.) Note the wider spacing between the lower voices for chords having a low average tessitura. Notated pitches represent complex tones rather than pure tones. *Source*: Huron (2001; see also Huron, 1993c).

widening of intervals in the bass is consistent with efforts to distribute the partials in a roughly uniform manner across the basilar membrane.

So why do composers do this? Of course we can't read the minds of musicians. But notice that this pattern of spacing is what you would expect to see if composers were attempting to minimize auditory masking. That is, spacing chords in this manner reduces the mutual interference between partials. Since the partials are easier to resolve, this means that it is easier for the auditory system to form clear auditory images of the individual notes or pitches. In short, arranging sonorities this way makes it easier for listeners to perceive the individual sound sources or musical parts.

Over the past couple of centuries, some music theorists have suggested that the tones within a chord are ideally spaced according to the harmonic series. Like the harmonic series, many chords have larger intervals separating the lower parts. However, the American composer Walter Piston recognized that musical practice doesn't conform to this formerly popular view.[7] If composers emulated the harmonic series when spacing chord members, then the spacing would be the same whether the chord was positioned low or high in overall tessitura. Instead, musical practice shows systematic changes of spacing with respect to register. A close-position triad sounds fine in the middle and upper registers but sounds muddy when played in the bass. Understanding how frequencies are mapped onto the basilar membrane provides a straightforward explanation for common musical practice.

In light of the research on auditory masking, we can formulate the following principle:

3. Minimum Masking Principle
Auditory masking for any vertical sonority is reduced when the sound energy is spread evenly across the basilar membrane. When typical harmonic complex tones are employed, an even spread of sound energy requires a wider spacing of tones as the sonority descends in register.

Incidentally, critical bands play a vital role in portable MP3 players. In order to fit many audio files onto a small storage device, the digital signal is compressed. The standard MPEG (MP3) compression method takes advantage of the relative insensitivity of the ear to multiple frequencies occurring within the region of the critical band.[8] MP3 players significantly reduce the amount of storage required for music files by throwing away sounds that would be masked in the human auditory system. In effect, MP3 players rely on the same principle as evident in the spacing of chords. For most listeners, the difference in sound quality between high-definition audio coding and an MP3 coding is negligible. But a dog or cat would immediately be able to detect the difference in sound quality: the MP3 compression method is optimized for the human cochlea, not the cochleas of cats or dogs.

It seems impressive that the esoteric experiments on critical bands that Harvey Fletcher carried out in the 1950s ultimately contributed to the design of music devices that people now commonly carry around in their pockets. However, the more impressive fact is that critical bands have influenced the organization of music for centuries. The tendency for composers to construct chords with wider intervals in the bass means that nearly every page of multipart musical notation has the human auditory system written all over it.

More about Masking

As noted at the beginning of the chapter, masking is not an all-or-nothing matter. We can have complete masking, various degrees of partial masking, or no masking at all. Figure 5.5 shows the masking effect of a single pure tone. The horizontal axis represents frequency, and the vertical axis represents sound intensity (in decibels). The "masker" tone is represented in the middle of the graph by the tallest vertical line; this masker tone has a frequency of 1,000 Hz and an 80 dB intensity level. The area of masking is represented by the triangular region beneath the inverted V, referred to as a *masking skirt*. Any sound that falls within the masking skirt is fully masked. To the left of the masker, the illustration shows a 950 Hz tone with an intensity level of 60 dB; it lies just under the skirt and so is completely masked. A tone that pokes up above the masking skirt will be partially masked. The illustration shows a 1,050 Hz tone with an intensity level of 70 dB. In partial masking, the masked sound is effectively reduced in loudness. Listeners can detect only the part that protrudes above the masking skirt. In effect, the 1,050 Hz tone is perceived as having an intensity of just 10 dB. Hearing scientists view the masker as raising the threshold of hearing for the partially masked sound.

All tones have the capacity to mask. In figure 5.5, both the 950 Hz and 1,050 Hz tones would have their own masking skirts; these have been omitted from the figure in order to avoid excessive clutter. Nevertheless, these tones would also have an effect on the 1,000 Hz tone, although the amount of masking would be less due to their lower intensity levels.

So far, we have been considering only the masking effect between pairs of pure tones. But in the real world, most sounds consist of many partials.

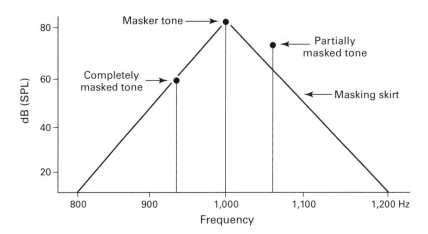

Figure 5.5
Masking effect of a 1,000 Hz pure tone with an 80 dB intensity level. Complete masking is represented by the triangular region under the masking skirt. The lower tone (950 Hz with an intensity level of 60 dB) is completely masked. The higher tone (1,050 Hz with an intensity level of 70 dB) is only partially masked. Asymmetrical spread of masking with frequency is not illustrated. After Egan and Hake (1950).

Auditory Masking

If we want to understand the aggregate masking effect of, say, a trumpet on a flute, then we must consider all of the partials produced by both instruments.

Rather than consider two different instruments, let's examine the simpler case when two identical complex tones differ only in pitch: that is, they have different fundamental frequencies. The two tones are equally energetic and have identical harmonic content. The first harmonics both have the same intensity, the second harmonics have the same intensity, and so on. In general, for complex tones, the energy tends to fade with successive partials. Although individual sounds vary considerably in timbre, the energy found in the upper partials is typically less than the energy found in the lower partials. This declining energy with increasing frequency is referred to as *spectral roll-off*. Figure 5.6 shows declining intensity for the first seven harmonics of a complex tone whose fundamental is 230 Hz. The vertical axis represents sound intensity or amplitude. Unlike the previous graphs, the horizontal axis in this case indicates position along the basilar membrane. This means that equal horizontal distances represent

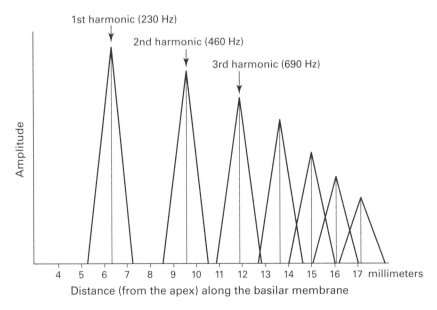

Figure 5.6
Masking skirts for seven harmonics of a complex tone with a fundamental frequency of 230 Hz. *Note*: When masking skirts are plotted with respect to position along the basilar membrane (rather than with respect to frequency), they are symmetrical.

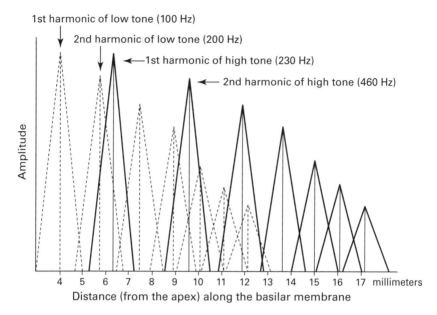

Figure 5.7
Masking skirts for two complex tones, both containing seven harmonics with identical intensities or amplitudes. The lower tone has a fundamental of 100 Hz (dashed lines); the upper tone has a fundamental of 230 Hz (solid lines). Significant mutual masking is evident. Notice that the lower partials of the higher tone mask the upper partials of the lower tone more than the other way around.

equal regions of potential masking. Remember that masking will occur only between partials that are within a millimeter of each other. In figure 5.6, masking skirts have been drawn for each of the harmonics shown.

Now consider the interaction of this tone with a 100 Hz complex tone having an identical harmonic recipe. In figure 5.7 the partials for both tones are shown. The partials of the lower tone are shown as dashed lines.

Notice that the upper partials of the lower-pitched tone are significantly lower in intensity than the neighboring partials of the higher tone. This means that the higher complex tone is at a significant advantage: the partials of the higher-pitched tone mask the partials of the lower-pitched tone more than the other way around.

The effect shown in figure 5.7 holds generally. For complex tones that exhibit spectral roll-off, the partials of higher-pitched tones will tend to mask the partials of lower-pitched tones more than the reverse. This means that more of the energy of the higher tone will escape being masked. All other things being equal, listeners will be able to detect higher-pitched tones better than concurrent lower-pitched tones, a phenomenon referred to as the *high-voice superiority effect*.[9]

In most of the world's cultures, there is a notable tendency to place the principal musical line or melody in the uppermost voice or part. This is true in the traditional musics of China, Japan, and Korea, of Indonesia and Southeast Asia, of India, of the Middle East and North Africa, and of Central and South America. In most Western four-part harmony, for example, the melody is commonly given to the soprano voice. Of course this tendency is not universal; in Western music, counterexamples include descant singing where an obbligato part is placed higher than the main melodic line. Another example occurs in barbershop quartets where the lead voice is positioned between the tenor and the baritone.[10] Similar exceptions can be found in non-Western music. Despite such exceptions, the general tendency to place the melody in the highest part can be observed all over the world.

On the face of it, there should be no impediment to placing melodies in the lowest voice or in some inner voice. Of course, any line can be made more noticeable simply by making it louder than the others. But when the concurrent parts are similar in loudness and timbre, the effect of relative pitch placement on masking becomes paramount. Although other factors might favor placing the melody in the uppermost voice, auditory masking provides a straightforward explanation for this common musical tendency.

Some perspective on the importance of this phenomenon can be gained if we consider the social circumstances in which much of the world's music-making has occurred. Throughout history and in nearly every documented culture, there has existed a pervasive prejudice against women. In many cultures, women's participation in music-making was actively discouraged, highly restricted, or even forbidden. In light of this prejudice, it is striking that so much of the multipart music of the past was organized to permit women to sing the foremost vocal part. Even when women were socially excluded from music-making, it is striking that young boys (also of comparatively low social status) still managed to command the principal melodic line. The implication is that the physiological structure of the basilar membrane helped to save women and children from what might have been a history of near-total musical exclusion.

In light of this research on masking, we can modify our minimum masking principle to include the following addition:

3. Minimum Masking Principle (continued)
    … *When common complex tones are used, higher-pitched tones will tend to mask lower-pitched tones more than vice versa.*

Auditory Irritation

Masking is a form of sensory interference where one sound gets in the way of another one. The same thing happens in vision. You might purchase cheap tickets to a popular concert only to discover that your view of the stage is obstructed by a pillar. Nearly everyone experiences visual obstruction as irritating or annoying, and for good reason: the obstruction makes it difficult to gather visual information from the environment. The presence of the pillar reduces the effectiveness of the visual system, and the irritation we feel motivates us to try to improve the situation. That is, the feeling of annoyance encourages us to try to restore the effectiveness of the visual system. We look around to see if there is an empty seat whose sightline isn't blocked by the pillar.

The situation is the same with auditory masking. When you are trying to carry on a telephone conversation, background sounds can be quite irritating. The feeling of irritation or annoyance motivates you to try to restore the effectiveness of your auditory system. You might move to a quieter space, or ask your neighbors to be quieter, or ask your conversation partner to speak louder, or stick your finger in your free ear. Notice that the irritation serves a useful purpose: if you didn't feel some annoyance or discomfort, you wouldn't be motivated to take steps to improve your ability to hear.

Suppose we had the opportunity to design our own auditory system. We've done our best to create a cochlea free of masking but can't get rid of masking entirely: there are physical limitations to how close two stimuli can be and still manage to keep them separated. We know that masking is potentially dangerous: if we can't hear the footsteps of some predator, we may end up on its dinner menu. So what else can we do to minimize the potentially onerous effects of masking? One useful addition is to make masking *irritating*. That is, whenever we recognize that masking is taking place, we generate negative feelings of annoyance or unpleasantness. The

Auditory Masking

feelings of discomfort will go away only if the masking goes away. With this kind of design, listeners will soon learn to behave in ways that ultimately reduce the masking: turn off the TV, close the door, ask people to be quiet, or simply leave a noisy environment. In other words, in trying to create an optimal hearing system, we can supplement the physiological design of the hearing mechanism with a behavioral component: a little bit of irritation can help preserve the effectiveness of the auditory system.

In order to make this behavioral component possible, the auditory system must somehow detect when masking is occurring. Consider the four situations shown in figure 5.8. In each example, only the masking skirt for the lower tone is shown, but the upper tone also produces a masking skirt so both tones are engaged in mutual masking. In the first case (figure 5.8a) the two pure tones are separated by more than 1 millimeter (more than a critical band). Consequently, no masking is produced: the auditory system is happily able to resolve both tones. In the second case (figure 5.8b) the tones have moved to within a critical band. Some masking is generated, but both tones are still audible despite the partial masking. In the third case (figure 5.8c), significant masking is occurring. Only a small bit of the higher tone is above the masked threshold, so the higher tone will be perceived very weakly. Finally, in figure 5.8d, complete masking has occurred. The higher tone has been entirely swallowed up by the masking effect of the lower tone. Consequently, the listener hears just the lower tone: the listener's experience would be the same if the higher tone were absent altogether.

Now suppose we are trying to design the optimal behavioral irritation for each of the four conditions illustrated in figure 5.8. In the first case, no masking occurs, so there is no need to generate any annoyance. In the

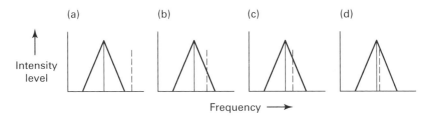

Figure 5.8
Four potential masking situations: (a) no masking, (b) moderate masking, (c) significant masking, (d) complete masking.

second case, slight masking occurs, so it would be appropriate to generate feelings of slight annoyance. In the third case, significant masking occurs, so we increase the sense of irritation, producing marked feelings of annoyance. In the final case, it would be appropriate to generate the greatest annoyance: however, there is a snag: the auditory system has no way of knowing that masking is occurring. The upper tone is completely masked and the cochlea behaves the same as if the higher tone were completely absent. We aren't justified in producing any feeling of annoyance in this case because we have no evidence that any masking has occurred. The only reasonable recourse is to assume that there is just one tone present in the environment—hence, no annoyance.

In the early 1960s, two seminal studies were carried out showing that the auditory system behaves in a way consistent with this scenario. Working in the Netherlands, Reinier Plomp and Willem Levelt measured the reported irritation or annoyance associated with different critical-band distances. They found that when two pure tones are separated by more than a critical band, listeners report no sensory irritation. When the distance is exactly a critical band, sensory irritation begins. As the two tones get closer together, the irritation increases until a maximum sensory irritation occurs when pure tones are separated by about 40 percent of a critical bandwidth.[11] As the two tones continue to get closer, the irritation decreases until it disappears entirely when the tones reach the same frequency (unison).

The work of Plomp and Levelt was independently replicated by Donald Greenwood. Then in the late 1960s, two young Japanese scientists, Akio Kameoka and Mamoru Kuriyagawa, repeated the experiments with Japanese listeners and found the same results. Kameoka and Kuriyagawa went on to devise the first mathematical model for calculating this sensory irritation for any arbitrary sound input.[12] In this discussion, we've been considering pairs of pure tones, but as we know, real-world sounds typically exhibit many partials. If a sonority contains (say) thirty partials, the amount of sensory irritation will depend on the distances between all of the successive partials—at least those partials separated by less than a critical band.

Consonance

In carrying out these experiments, different researchers asked their listeners slightly different questions: How irritating do these frequency pairs sound?

How annoying is this sound? How ugly is this interval? How nice does this sonority sound? And so on. These sorts of questions might sound familiar. They are similar to questions that music scholars have been asking for centuries: How dissonant is this sound? How consonant is this interval?

Since about 1880 or so, many experiments have been carried out investigating what musicians call consonance and dissonance. These experiments produced confusing, sometimes even conflicting results. Often the aim of these experiments was to determine the relative consonance for different harmonic intervals. Nearly all experiments showed that seconds and sevenths were judged more dissonant than thirds and sixths. Moreover, nearly all listeners judged minor seconds more dissonant than major seconds and major sevenths more dissonant than minor sevenths. But the experiments often produced different results for the other intervals. For example, some experiments found that perfect fifths were judged more consonant than major thirds, while other experiments showed the reverse results.

Some of the discrepancy appears to lie in the specific instructions given to listeners. The results can differ depending on whether we ask listeners to judge *consonance*, or *pleasantness, smoothness, harmoniousness, euphoniousness*, or something else. These words don't necessarily mean the same thing, and listeners may interpret the same word differently. Nor can we assume that words like *dissonant, roughness, disphonious, unpleasant, irritating, annoying, ugly*, and so on mean the same thing. In 1962, Van de Geer, Levelt, and Plomp carried out an important study comparing different descriptive labels for "consonant" sounds. Their research suggests that several different phenomena are going on at the same time.[13]

Another reason that the experiments produced diverging results can be traced to the sounds used in different experiments. Experiments have variously used tuning forks, organ pipes, plucked strings, pianos, electronically generated pure tones, and even sirens. Some experiments examined only intervals smaller than an octave; others looked at intervals spanning a larger range. Some experiments used only tones in the middle register, whereas others used tones in higher or lower registers. Today we know that consonance is influenced by many auditory and cultural factors.

Taken at face value, the existing research literature suggests that at least ten factors influence the perceived pleasantness of a sonority. For example, culture is known to play a role: people prefer sonorities that are culturally familiar. Musicians are known to respond differently from nonmusicians.

There are also individual differences. Among the low-level auditory factors, the perceived pleasantness of a sonority is known to be influenced by the loudness of the tones, the timbre of the tones, where they are positioned in absolute register, as well as the specific tuning used. At this point, readers will not be surprised to learn that among the factors known to influence judgments of pleasantness is the critical-band distance separating successive partials in the sonority. The greatest irritation is evoked when a sonority contains several pairs of partials whose separation along the basilar membrane approaches the maximum associated with 0.4 millimeters.[14] Moreover, many of the discrepancies observed between experiments using different timbres and different registers can be accounted for by the mechanics of the basilar membrane.

Once again, the research suggests that what we have been calling "sensory irritation" is just one of many factors influencing the reported pleasantness of a sonority. Moreover, we should be careful to avoid equating "irritation" with "dissonance." From a biological perspective, feelings of irritation are bad. Like all bad feelings, auditory irritation is intended to get us to behave in ways that preserve us from harm. But biological goals differ from musical goals. There are times when irritation can prove musically useful. The best music is not music that avoids dissonance.

Having cautioned ourselves against the idea that dissonance can simply be reduced to sensory irritation and having acknowledged that consonance is a complex phenomenon that involves cultural and other contextual factors, and having reminded ourselves that biological and musical goals are not the same, let's press ahead and examine musical practice in light of the experimental data collected on consonance and dissonance. For the remainder of this chapter, we'll avoid speculation about the origin or cause of consonance and dissonance. Instead, we'll simply treat consonance and dissonance as an empirical observation: some Western-enculturated listeners report that some vertical sonorities in some circumstances sound more consonant, euphonious, or pleasant (for example), than others.

Figure 5.9 shows the results from careful experiments carried out by Georg Kaestner at the University of Leipzig in the early twentieth century.[15] Kaestner played different harmonic intervals to German listeners and asked them to judge their "consonance" (*Konsonanz*). For musicians steeped in traditional Western theory, some of the perceptual results might seem odd: nonmusician listeners typically judge major and minor thirds and major

Auditory Masking

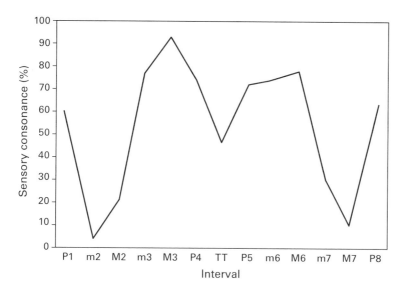

Figure 5.9
Reported consonance for complex tones from Kaestner (1909).

sixths as sounding more pleasant or consonant than perfect fourths and fifths, and fourths and fifths are reported as sounding more consonant than perfect unisons and octaves. Other results are consistent with traditional theory: the most dissonant intervals are seconds and sevenths, with minor seconds and major sevenths judged the most dissonant. The tritone (TT) also exhibits a fairly low consonance.

Let's begin with an incredibly naive hypothesis: that composers aim to give listeners lots of the "nice"-sounding consonances and try to avoid the ostensibly "annoying" dissonances. To test this hypothesis, we might simply tally up how often each harmonic interval occurs in a sample of some music.

Figure 5.10 plots such a distribution for a sample of music by J. S. Bach.[16] The plotted line indicates the percentage of occurrence for harmonic intervals ranging in size from the unison to the octave. The bars reproduce Kaestner's consonance data shown in figure 5.9. The two sets of data are broadly similar. For example, there are very few minor seconds and major sevenths, and thirds and sixths predominate. This seems to imply that Bach uses the various intervals in proportion to their degree of consonance. However, there are some notable discrepancies.

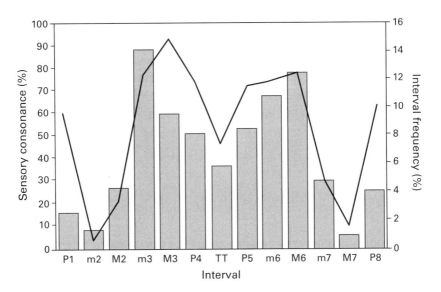

Figure 5.10
Comparison of sensory consonance for complex tones (line) from Kaestner (1909) with interval prevalence (bars) from a sample of music by J. S. Bach. Notice especially the discrepancies for P1 and P8. Source: Adapted from Huron (1991b).

One discrepancy is that the minor and major thirds appear to be switched: the consonance data would predict that major thirds would be more prevalent than minor thirds. A possible explanation for this can be found in the interval content of scales and common chord types. Both the major scale and the harmonic minor scale permit the creation of four minor thirds but just three major thirds. This imbalance is echoed in the organization of common chords. Major and minor triads both contain one major third and one minor third. However, the diminished triad (two minor thirds) is much more commonly used than the augmented triad (two major thirds)—and both the dominant seventh chord and the minor-minor-seventh chord contain one major third and two minor thirds. So the greater prevalence of minor thirds over major thirds might simply reflect the interval content of scales and chords. The relative prevalence of major and minor sixths (inversions of thirds) is also consistent with this suggestion.

The biggest discrepancies in figure 5.9 are evident for the unison (P1) and octave (P8) intervals. Unison and octave intervals are not common in Bach's writing, even though nonmusician listeners judge them as sounding fairly "consonant." What's going on?

Recall that harmonic fusion is the tendency for two or more complex tones to be perceived as forming a single auditory image. This might occur, for example, when two harmonic tones form an octave. In the late nineteenth century, the German psychologist Carl Stumpf suggested that harmonic fusion might be the cause of consonance. Stumpf suggested that when two complex tones fuse together into a single auditory image, they are more likely to be judged as producing a pleasant or euphonious sound. In carrying out further experiments, however, Stumpf realized that his idea must be wrong. Although Stumpf abandoned his original view, his suggestion nevertheless remained popular. Today we know that Stumpf was right to reject his original theory. Although harmonic fusion may influence judgments of consonance, harmonic fusion and consonance are different perceptual phenomena. Albert Bregman likes to summarize the confusion between consonance and harmonic fusion as a confusion between "smooth sounding" and "sounding as one."[17] If we use Bregman's terms, intervals like unisons and octaves sound "smooth" but also tend to sound "as one." Intervals like major thirds and sixths sound smooth but don't tend to sound as one. Intervals like tritones don't sound smooth or sound as one.

Suppose for the moment that Bach was aiming to produce a "smooth" sound, but also wanted to avoid situations where concurrent parts inadvertently sound "as one." What would the graph of interval prevalence look like? We'd expect that graph to show evidence of both factors. First, we would expect fused harmonic intervals to be avoided in proportion to the strength with which each interval promotes harmonic fusion. That is, unisons would be expected to occur less frequently than octaves, which would occur less frequently than perfect twelfths, which would occur less frequently than other intervals. Second, we would expect to see intervals in proportion to the reported "consonance," "smoothness," or "pleasantness." In a study of interval use in the music of J. S. Bach, I found exactly this pattern in a sample of his polyphonic works.[18] Of course harmonic (concurrent) octaves and fifths occur regularly in music, but, remarkably, they occur less frequently in polyphonic music than they would in a purely random juxtaposition of voices. That is, if you shift two musical parts with respect to each other by some random duration (say, shifting the entire treble part ahead by seventeen quarter durations), the resulting nonsense score will typically contain more harmonic octaves and fifths than in the original score. Note that this observation is independent of the avoidance of parallel unisons, fifths, or octaves. As simple static harmonic intervals, these intervals are actively minimized in Bach's polyphonic works. Considering the importance of octaves and fifths in the construction of common Western chords, the fact that Bach minimizes their presence is striking.[19]

As we've already seen, figure 5.10 seems to support this two-factor interpretation of Bach's harmonic intervals. But we can do better than relying on impressions. There is a statistical technique (called *multiple regression* analysis) that allows a more precise way of quantifying the influence of two or more factors on some phenomenon. By way of illustration, suppose we were interested in possible factors contributing to a person's height. Height is probably influenced by nutrition, especially the number of calories consumed when a person is growing up. But height also seems related to genetic inheritance: tall people often have tall relatives. For a large group of people, suppose that we have individual measurements of their heights, measures of their average caloric intake as children, and measures of the heights of their parents. Multiple regression analysis will tell us two important things: (1) which of the factors (childhood nutrition or parental height) contribute more to predicting a person's height and (2) how well (or how poorly) we can predict a person's height using just the factors being examined. That is, multiple regression analysis can tell us whether there are other factors (that we haven't identified) that contribute more to a person's height. The accuracy of prediction is represented by a statistical value designated R^2; an R^2 of 1.00 means a perfect ability to predict someone's height from these simple variables. An R^2 of 0.50 means you can account for half of the variability: that is, you've identified factors that account for half of the variance in people's heights. An R^2 of 0.00 means that you have nothing: the variables you've identified have no bearing on a person's height. A hypothetical analysis might show that childhood nutrition accounts for 50 percent of the variability and parental height accounts for another 20 percent. Together these factors account for 70 percent of the variability in people's heights, but there remains another 30 percent of the variability that is not related to these two variables.

Returning to Bach's music, we can use Kaestner's consonance data along with Stumpf's data for harmonic fusion to try to predict the frequency of occurrence of different intervals from the unison to the octave. In my study, the calculated multiple regression for both factors resulted in an R^2 of 0.88. This is consistent with the interpretation that nearly 90 percent of the variability in Bach's harmonic interval use can be attributed to the twin compositional goals of the pursuit of consonance and the avoidance of harmonic fusion.[20] The multiple regression analysis also suggests that Bach pursued both of these goals with approximately equal resolve or success. Although this study does not prove anything regarding Bach's compositional intentions, the results are consistent with the idea that Bach preferred harmonic intervals in proportion to the degree to which they promote consonance and in inverse proportion to the degree to which they promote fusion. Using Bregman's terminology, the results imply that Bach aimed to produce contrapuntal combinations that sound "smooth" without the danger of "sounding as one." (In chapter 15, we will consider possible aesthetic motivations for this practice.)

As a final observation, notice that the experimental results pertaining to consonance and harmonic fusion might be used to illuminate the traditional classification of harmonic intervals. Music theorists traditionally distinguish three types of harmonic intervals: *perfect consonances* (such as perfect unisons, octaves, fourths, and fifths), *imperfect consonances* (such as major and minor thirds and sixths), and *dissonances* (such as major and minor seconds and sevenths, and tritones). These three types of intervals can also be distinguished using the criteria of consonance and harmonic fusion. Perfect consonances typically exhibit high consonance and high harmonic fusion. Imperfect consonances have high consonance and comparatively low harmonic fusion. Dissonances exhibit low consonance and low harmonic fusion. (There are no equally tempered intervals that exhibit low consonance and high harmonic fusion although the effect can sometimes be produced using grossly mistuned unisons, octaves, or fifths.)

Reprise

This chapter began by discussing the phenomenon of auditory masking, that is, the ability of one sound to obscure another. We discovered that masking arises from mutual interference of pure tones along the basilar membrane. When two (pure) tones evoke points of activity that are within about 1 millimeter, they begin to mask each other. When mapped against a musical scale, the distance of 1 millimeter represents a much larger interval in the bass than in the treble region. Above middle C, the region of interference ("critical band") is roughly a minor third in size. Below middle

C, regions of interference get larger as you go lower in pitch. We saw that composers usually space tones in chords in a way that is consistent with a relatively even spreading out of energy along the organ of hearing. This is achieved when the distance between chordal tones is progressively larger as the pitch descends. We also saw how masking accounts for the high-voice superiority effect. That is, for complex tones exhibiting spectral roll-off, higher-pitched (complex) tones generally mask lower-pitched (complex) tones more than the other way around. The upper-most tone in a chord is typically the tone that is least masked—and so it is the most easily heard.

Since masking reduces the effectiveness of the auditory system, it also tends to lead to auditory irritation or annoyance. These mildly negative feelings are intended to encourage listeners to take actions that might restore the ear's sensitivity. Although other factors are involved (including enculturation), this annoyance is a contributing factor to the multifaceted experience commonly referred to as dissonance. Finally, we saw that the distribution of harmonic intervals in music by J. S. Bach is consistent with reducing dissonance while simultaneously minimizing harmonically fused intervals.

6 Connecting the Dots

In chapter 3 we introduced the concept of an auditory stream, the subjective sense of sonic line or singular sound activity continuing over time. As we saw, streams can arise even when the sound events are intermittent. For example, we hear a continuous ticking clock even though the individual ticks are physically isolated. We also learned that a sense of unitary sonic line can sometimes emerge when the successive sounds are produced by different sound sources. For example, the sound of someone walking is actually the sound of two independent acoustic generators—the sounds produced by the left and right feet. I once had a shoe that produced a high-pitched squeak, and for several days my footsteps did indeed sound like two streams—something like the alternation of a bass drum and a squeeze toy. But this situation is unusual. With footsteps, we don't usually hear some kind of *hocket*—two distinct sound sources taking turns. Instead, we usually hear a single line of sound with no distinction between the left and right feet.

Like footsteps, many musical instruments have multiple sound generators—often one or more generators for each pitch. Examples include the piano, organ, marimba, chimes, accordion, harmonica, and dulcimer (there are lots of others). In chapter 3, we discussed a harp playing a scale: amazingly, the ear connects the dots and so we tend to hear the eight successive tones as forming a single line of sound rather than eight individual sound sources taking turns. Because we experience this every day, musicians don't find this especially noteworthy or mysterious. But hearing scientists are not so complacent; they understand that creating a single auditory stream from multiple independent sound sources is pure magic. We can visually glimpse some of this magic when watching a handbell choir. Who would have thought that a smooth melody could emerge from a chaotic flurry of hand waving? In connecting the sonic dots, the brain is doing something very sophisticated—and something we take entirely for granted.

In this chapter we address the question of how successive sounds can give the impression of being connected together. Instead of focusing on the formation of static auditory images, we will focus on the formation of continuous auditory streams. This perception of connectedness is crucial for music perception because if we heard only independent tones, music would be incapable of conveying a sense of line or figure; melody would be impossible. In addition, we address how connecting sounds together might relate to the perception of motion or movement. We will see that musical melodies are consistent with a fundamental law of muscle-powered movement, and that complying with this law contributes to the perception of motion. Finally, we will see how coordinated movement between simultaneously lines influences stream perceptions.

Continuity

So how do successive sounds link together to form a musical line or auditory stream? A good place to begin is with the most obvious factor: temporal continuity. It is easier to connect dots when the dots are elongated into dashes, and connecting the dashes together is easier yet when the dashes are lengthened so they just touch each other. When sounds are sustained so that the end of one sound coincides with the start of the next, the impression of connectedness becomes compelling.

As a research question, our attention will necessarily focus on the borderline cases. When sounds are short—when they don't sustain from one sound to the next—how is it possible for us to sometimes hear them as connected? How can a sequence of staccato notes form a melody?

Even after a sound has been physically extinguished, it nevertheless lingers in consciousness. A good illustration can be found in a phenomenon called *auditory induction*, where listeners can be induced to hear sounds that are in fact absent. The phenomenon is robust; anyone can experience it, and you don't need to take any mind-altering drugs.

At the University of Wisconsin–Milwaukee, Richard Warren and his colleagues generated patterns of sounds in which a faint tone alternated repeatedly with a loud noise (see figure 6.1). The faint tone stopped at the same moment that the loud noise started, and vice versa. But this is not

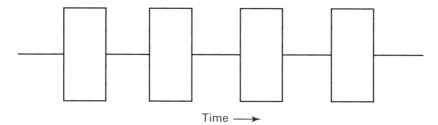

Figure 6.1
Schematic illustration of *auditory induction*. A faint tone (line) alternates with a series of noise bursts (blocks). Listeners report hearing the faint tone as continuing throughout the noise bursts even though the tone is physically absent.

what listeners heard. Instead, they heard a faint tone that was continuous throughout, with periodic bursts of noise superimposed over this steady tone. Although the tone is physical absent during the noise bursts, listeners insist they can hear it.[1]

This effect doesn't occur for all tones. It occurs only if the faint tone lies within a critical band of the noise burst. If the tone and noise are more than a critical band apart, the listener will accurately hear a "beeping" faint tone alternating with the noise bursts. It is only when the tone and the noise stimulate the same region of the basilar membrane that listeners hear the tone as continuous. If the tone were truly present throughout the noise burst, it would have been entirely masked and the auditory system could not possibly have detected its presence. It is this potential for masking that accounts for the hallucination. In effect, the brain detects a tone; then it detects a noise (that would be expected to mask the tone if the tone were still present); then the noise disappears, and the tone is detected again. Under the circumstances, it is not unreasonable for the auditory system to assume that the tone sounded continuously throughout the noise. What is amazing is that this assumption gets translated into an auditory image: listeners actually hear the pitch of a (nonexistent) sound continuing throughout the noise. In chapter 4 we noted that researchers are convinced that pitch is an auditory phenomenon (something constructed in your head), not an acoustic phenomenon (something "out there" in the world). Auditory induction is just one of many phenomena that illustrate the purely mental origin of pitch.

In effect, auditory induction mentally reinstates sounds that the auditory system would expect to be masked, even when the sounds are truly absent. Warren and his colleagues were able to document robust auditory induction effects for both pure tones and noises. With pure tones, listeners can commonly be coaxed into hearing nonexistent tones lasting up to a third of a second in duration. With bands of noise, auditory induction may be achieved for impressive durations of 20 seconds or more. That is, under the right conditions, listeners can "hear" a sound persisting for 20 seconds even when the sound is entirely absent.

Auditory induction depends on the presence of auditory masking. In an unaccompanied melody, masking will be slight or negligible, so auditory induction isn't especially germane. The important lesson from auditory induction is that auditory images don't necessarily end when the sound ceases; the images can persist well beyond the sound's physical duration.

With this lesson in mind, let's now consider a musically more pertinent phenomenon, one that comes from research on auditory memory. When a sound ends, the auditory system will retain a sort of afterimage that soon fades away. Like the flash of a visual image, auditory images tend to linger briefly after the physical cessation of the stimulus. Cornell University psychologist Ulric Neisser dubbed this phenomenon *echoic memory*.[2] Various experiments have measured or estimated the duration of echoic memory. The best measurements suggest that echoic memory has a half-life of about 1 second; that is, after about 1 second, the vividness of the auditory image is roughly halved.[3]

A useful question to ask is: What is the longest silent gap between two tones that nevertheless allows listeners to hear the tones as connected? An experiment carried out by Dutch researcher Leon van Noorden suggests that the longest silence is roughly 800 milliseconds—about eight-tenths of a second.[4] This duration is similar to measurements of echoic memory, and so most researchers think that echoic memory provides the bridge between two successive sound images. It looks as if the auditory afterimage preserved by echoic memory helps the auditory system to connect the dots.

In general, the research suggests that clear auditory streams are best evoked by sounds that abut one another in time or are separated by only brief silences. Accordingly, we might formulate the following principle:

4. Continuity Principle

The perception of an auditory stream is facilitated when using contiguous rather than intermittent sound sources. Intermittent sounds are less likely to be perceived as forming a continuous stream when silent gaps exceed about 800 milliseconds.

Continuity and Musical Practice

The musical implications of this principle are readily apparent. In the first instance, the continuity principle is evident in the types of sounds commonly used in music-making. Compared with most natural sound-producing objects, musical instruments are typically constructed so as to maximize the period of *sustain*. Most of the instruments of the Western orchestra, for example, are either blown or rubbed—modes of excitation that permit relatively long-lasting or continuous sounds. Even in the case of percussion instruments like the piano and the vibraphone, the history of the manufacture of these instruments shows a marked trend toward extending the resonant durations of the sounds produced. For example, over the course of history, pianos became progressively heavier, with large metal frames that enabled increased string tension, resulting in tones of longer duration. Similar historical developments can be traced for such instruments as timpani, gongs, and marimbas. In some cases, musicians have gone to extraordinary lengths in order to maintain a continuous sound output. In the case of wind instruments, the disruptions due to breathing have been overcome by such devices as bagpipes, mechanical blowers (as in pipe organs), and devices that respond to both "inhaling" and "exhaling" (e.g., the accordion and perhaps the harmonica). In some cultures, performance practices exist whose sole purpose is to maintain an uninterrupted sound. A particularly impressive example is the practice of circular breathing as used, for example, in the Arab shawm (*zurna*) and the Australian didgeridoo. Here the performer's cheeks are used like the bag of a bagpipe: when the performer inhales (quickly through the nose), the cheeks are squeezed so that the airstream is never interrupted. For stringed instruments, bowed modes of excitation have been widespread. In the extreme, continuous (circular) bowing mechanisms have been devised such as used in the hurdy-gurdy. In the case of the guitar, solid-body construction and controlled electronic feedback became popular methods of increasing the sustain of plucked strings.

Apart from the continuous character of most musical sounds, composers tend to assemble successions of tones in a way that suggests a persistent or ongoing existence. A notable (if seemingly trivial) fact is that the majority of notated tones in music are followed immediately by a subsequent tone. For example, roughly 93 percent of all tones in vocal melodies are followed immediately by another tone. The corresponding percentage for instrumental melodies exceeds 98 percent.[5] The exceptions to this observation are themselves telling. Successions of tones in vocal music, for example, are periodically interrupted by rest periods that allow the singer time to breathe. In most of the world's music, the duration of these rest periods is just sufficient to allow enough time to inhale (about 1 or 2 seconds). (Longer rests typically occur only in the case of accompanied vocal music.) Moreover, when music-making does not involve the lungs, pitch successions tend to have even shorter and fewer interruptions. Once again, as a general observation, we can note that in most of the world's music-making, there is a marked tendency to maintain a more or less continuous succession of sound events.

In tandem with efforts to maintain continuous sound outputs, musical practices also reveal efforts to terminate overlapping sounds. When musicians connect successions of pitches, problems can arise when each pitch is produced by a different vibrator. The dampers of the piano, for example, are used to truncate each tone's natural decay. When playing a scale, for example, pianists lift one finger at the same time that the next finger depresses the ensuing key. The dampers terminate one tone at roughly the same moment that the hammer engages the next tone. This practice is not limited to the piano or to Western music. Guitar players frequently use the palm of the hand to dampen one or more strings—and so contribute to the sense of melodic continuation when switching from string to string. In the case of Indonesian gamelan music, proper performance requires the left hand to dampen the current resonator while the right hand strikes the next resonator, a technique known as *tutupan*. In all of these examples, the damping of physical vibrators suggests that musicians are trying to create the illusion of a single continuous acoustic activity.

Of course music-making also entails the use of brief sounds, such as sounds arising from various percussion instruments like the wood block. However, musicians tend to treat such brief sounds differently. Brief sounds are less likely to be used to construct lines of sound such as melodies. When instruments with rapid decays are used to perform melodies, they often employ tremolo or multiple repeated attacks, as in music for xylophone or steel drums. When brief tones are produced by nonpercussion instruments (e.g., staccato), there is a marked tendency to increase the rate of successive tones.

Intrigued by this idea, my colleague Randolph Johnson studied performances of the same works played on guitar and banjo.[6] Since banjo strings decay faster than guitar strings, Johnson reasoned that there might be a tendency for the banjo performances to be faster than guitar performances of the same pieces. He located pairs of recordings for a number of works, explicitly limiting his study to purely instrumental works so the tempos would not be influenced by vocal demands. Johnson found no significant differences in the tempos between the two sets of recordings: beat-for-beat, the banjo renditions were no faster or slower on average than the guitar renditions. However, he found a big difference in the *note rate*—the number of notes played per second. The banjo renditions commonly played twice as many notes per beat as the guitar versions. If a performer wants to create the impression of a continuous sound texture, then it makes sense: when tones have shorter durations, consider cramming more tones into the same period of time.

Although there are exceptions, in most of the world's music-making, there is a marked preference for some sort of continuous sound activity; moreover, when instrumental sounds are brief in duration, such sounds are often assigned to nonmelodic musical tasks or are performed with more rapid successions of notes.

Tones in Motion

In 1950, two young psychologists carried out a simple experiment involving back-and-forth alternations between two tones. The two psychologists were George Miller (who was later a recipient of the U.S. National Medal of Science) and George Heise (who later became famous for discovering the antianxiety drug Valium). Miller and Heise observed that trill-like alternating pitches can produce two different perceptual effects. When the tones are close in pitch, alternations evoke a sort of undulating effect, like a single wavering line. However, when the pitch separation is large and rapid, the perceptual effect becomes two static beeping tones with no sense of movement between them.[7] The effect is illustrated musically in figure 6.2.

a)

b)

Figure 6.2
Effect of pitch separation on the sense of auditory motion. (a) Neighboring pitches evoke the perception of a single undulating line. (b) Distant pitches evoke the perception of two static beeping tones with no sense of movement between the pitches.

The trill notated in figure 6.2a sounds like a single wavering line, but the tremolo notated in figure 6.2b sounds more like two stuttering pitches with no movement connecting them. Although the term *auditory stream* wasn't coined for another two decades, Miller and Heise had discovered that pitch sequences could either cohere into a single stream or break apart into two or more streams.

Miller and Heise measured the amount of pitch separation required to ensure the perception of a single line. They found that when the interval between the alternating tones is two semitones or smaller, a single undulating line is nearly always heard.

For musicians, this discovery seems obvious. For centuries, the Swiss (and others) have been yodeling, in which a rapid alternation between widely separated pitches allows the singer to produce a sort of virtual harmony. Since at least the Baroque period, composers have written music with fast pitch jumps that conveys the impression of more than one concurrent line. Music theorists have referred to the phenomenon as *pseudo-polyphony* or *compound melodic line*. Although the effect has long been a staple of musical composition, the psychologists who became captivated by this phenomenon did a splendid job of studying it in detail.

As often happens in science, several different researchers stumbled across this same phenomenon without any knowledge of each other's work. Although Miller and Heise were the first to publish a description of the effect, similar observations were published by researchers working independently in Canada, Italy, the Netherlands, and the United States.[8]

Connecting the Dots

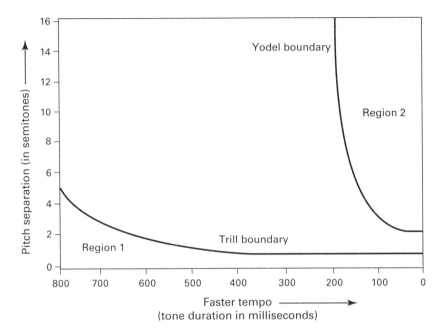

Figure 6.3
Influence of interval size and tone duration on the perception of alternating tones. In region 1, listeners always hear one stream (small interval sizes and slow tempos). In region 2, listeners always hear two streams (large interval sizes and fast tempos). In the large region between the two boundaries listeners may hear either one or two streams depending on the context and the listener's mental disposition. *Source*: After van Noorden (1975, p. 15).

Miller and Heise observed that whether a listener hears one line or two depends on two factors: the pitch distance between the two tones and the speed of alternation. In 1975, Dutch researcher Leon van Noorden expanded and refined these observations. He carefully mapped the effect of tempo and pitch separation on whether listeners hear one or two lines of sound. He discovered that the perception of two alternating pitches can produce one of three effects, as illustrated in figure 6.3. The horizontal axis in figure 6.3 is the speed of alternation, expressed in the duration of each tone in milliseconds. Faster tempo requires shorter tone durations (so the numbers get smaller as you move to the right). The vertical axis for figure 6.3 indicates the interval separating the two pitches (in semitones): the intervals are larger as you move upward. When the two pitches are very close together, the resulting sequence is always perceived as a single stream.

This area is indicated in figure 6.3 as Region 1, below the *trill boundary* (lower line). Conversely, when the pitch distances are large or the tempo is fast (or both), two streams are always perceived. This situation is indicated as Region 2, to the left of the *yodel boundary*.[9] Van Noorden also identified a very large middle region (between the two boundaries), where listeners may hear either one or two streams depending on the context and the listener's mental disposition. In some circumstances, listeners can even choose to hear either one stream or two streams. (We'll discuss this phenomenon further in chapter 13.)

Pitch Proximity

Notice that the slope of the trill boundary is relatively flat, but the slope of the yodel boundary is nearly vertical. The existence of two boundaries (rather than one) suggests that pitch-based streaming involves at least two different perceptual principles. Let's begin by focusing on the relatively flat trill boundary. Below this boundary, the perception of a single auditory stream is assured. So if a composer employs pitch successions that lie within this region, listeners will always hear the pitches as connected together in time. In 1967, a Harvard University graduate student in psychology, Jay Dowling, set out to connect Miller and Heise's observations to musical practice. Dowling was the first to suggest that a musical melody might be a kind of auditory stream; that is, melodies are sequences of tones that connect together perceptually as a single sonic line.[10] Part of what makes a melody a melody is that we hear successive tones as connected in time rather than as independent pitches.

Dowling tallied the melodic intervals used in a sample of music and formally demonstrated what musicians would regard as obvious: in nearly all melodies, small intervals predominate—intervals that fall within the trill boundary and so guarantee a sense of connectedness. Dowling also examined Baroque solo passages that listeners had identified as evoking the perception of multiple concurrent musical lines. In these passages, large intervals predominate. In a sample of pseudo-polyphonic passages, Dowling found that the composers never use intervals less than the trill boundary.[11]

Music theorists traditionally distinguish two kinds of melodic interval: step (or *conjunct*) motions and leap (or *disjunct*) motions. In Western music, the dividing line between step and leap motion is traditionally placed between a major second and a minor third; that is, a major second (two semitones) is considered a step, whereas a minor third (three semitones) is considered the smallest leap. In some cultures, such as those that employ the common pentatonic scale, the maximum step size is roughly three semitones. For tones longer than about 200 milliseconds (1/5 of a second), three-semitone intervals are still most likely to result in the perception of sequential connection. It seems possible that the trill boundary in effect identifies a perceptual basis for the traditional distinction between *conjunct* and *disjunct* melodic motions. The use of conjunct intervals will typically ensure that listeners hear a single auditory stream—a single connected line of sound.

In light of the research on pitch-based streaming, we can state the following principle:

5. Pitch Proximity Principle

The perception of an auditory stream can be ensured by close pitch proximity between successive tones. The perception of a single stream is almost certain when pitch movement is within the trill boundary. Intervals of two semitones or less will normally ensure the perception of a single auditory stream, regardless of tone duration.

For centuries, people have observed that successive pitches in melodies tend to be close together.[12] That is, most melodic intervals conform to the pitch proximity principle. This is illustrated in figure 6.4, which plots the distribution of interval sizes for melodies from a number of cultures: American, Chinese, English, German, Hasidic, Japanese, and sub-Saharan African (Pondo, Venda, Xhosa, and Zulu). In most of the world's music, there is a preponderance of small intervals.

It turns out that pitches compete with each other to capture or possess subsequent pitches as part of a stream. The pitch proximity principle helps us understand this process. Such stream competition is illustrated in figure 6.5 where two concurrent pitches (D4 and C5) are followed by a single pitch (A4). With regard to streaming, there are four logical possibilities, not all of which are equally likely for listeners. The A might be captured by the pitch C, so that the lower part ends with a rest (example b). Conversely, the pitch A might be captured by the D, so that the upper stream ends (example c). Alternatively, both parts might be perceived as continuing with the A

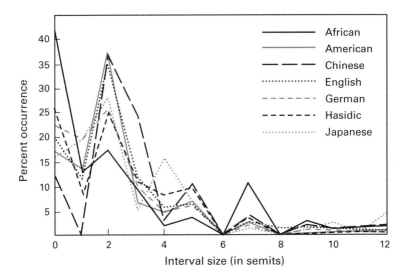

Figure 6.4
Frequency of occurrence of melodic intervals in notated sources for folk and popular melodies from several cultures. Interval sizes have been rounded to the nearest equally tempered semitone. *Source*: Huron (2001).

Figure 6.5
Illustration of pitch competition. Four auditory interpretations are illustrated for the stimulus shown in example (a). Ensuing pitches commonly stream to the nearest preceding pitch, as in example (b). Other stream organizations (c–e) are less likely to be heard.

(i.e., unison). Finally, it is possible that the auditory system will hear the two initial streams as ending, with a new (third) stream beginning with the onset of the A. Musical intuition suggests that the nearest stream will tend to win out in such competitions, and that's exactly what the experimental research shows. Since the A is closer to the C than to the D, listeners will most likely hear the A as connected to the C. In general, ensuing pitches tend to connect to the nearest previous pitch.

One of the most striking demonstrations of the role of stream competition in music can be seen in the case of crossing parts. Suppose that we

Connecting the Dots 75

Figure 6.6
Illustration of part-crossing. Listeners tend to hear the two lines as bouncing away from each other (b) rather than crossing (a).

have two concurrent musical parts. Each part will best connect to the nearest subsequent pitch. Now suppose we would like the upper part to cross beneath the lower part. Can we do this while ensuring that each part moves to the nearest subsequent pitch? *No.* Except in the case where both parts move to the same pitch (unison), the crossing of parts with respect to pitch will always violate the pitch proximity principle. Since unison intervals promote tonal fusion, it is impossible for simultaneous parts to cross without some disruption of the auditory streaming.

So what happens when the pitches of two sound sources actually do cross? At the University of California, San Diego, Diana Deutsch carried out the pertinent experiments.[13] She played concurrent ascending and descending tone sequences using tones having identical timbres. (See figure 6.6.) Two possible perceptions of intersecting pitch trajectories are shown in figure 6.7. The first illustration (crossed trajectories on the left) represents the simpler pitch contour. What could be simpler than one line ascending and a second line descending? Yet Deutsch found that the lines are perceived to switch direction at the point where their trajectories cross. Listeners hear the lines as bouncing away from each other rather than crossing.[14]

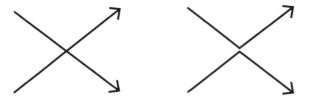

Figure 6.7
Schematic illustration of two possible perceptions of intersecting pitch trajectories. Bounced perceptions (right) are more common for stimuli consisting of discrete pitch sequences, when the timbres are identical.

Figure 6.8
Illustration of rhythmic discontinuity caused by crossing parts. When passage (a) is performed, listeners have a strong tendency to hear passage (b). The upper stream becomes syncopated at the point of pitch crossover, whereas the lower stream ceases to be syncopated at the point of crossover.

An even more interesting demonstration of the failure of crossing parts is evident in another one of Deutsch's experiments (illustrated in figure 6.8). Here Deutsch alternated the pitches for ascending and descending musical lines. The tones in the descending lines are "on the beat" while the tones in the ascending lines are "off the beat." At the point where the lines cross, a "bounced" percept would cause a disruption of the rhythm. If listeners hear the streams as changing directions, then the high stream tones would now be off the beat, while the low stream tones would be on the beat. That is, if pitch proximity is the dominant factor, then listeners would tend to hear the high tones as belonging to a single voice ("the high voice") even if that means that the rhythmic pattern becomes incoherent. And this is precisely what listeners report hearing. In other words, grouping all the high pitches together (and grouping all the low pitches together) is more important than preserving a consistent rhythmic pattern. In this demonstration, pitch-based streaming takes precedent over rhythmic regularity.

Given the work on stream competition for pitches, we can revise the pitch proximity principle accordingly:

5. Pitch Proximity Principle (continued)
… *All other factors being equal, when more than one stream is present, subsequent pitches will tend to be captured by the nearest existing stream.*

The effect of stream competition is amply evident in musical practice. In part-writing, composers have sometimes gone to extraordinary lengths to avoid the inadvertent crossing of parts. In a tediously detailed study of part-crossing, I formally tested three competing hypotheses regarding how J. S. Bach maintained voice separation with respect to pitch.[15] For example, one interpretation is that Bach simply assigned different tessituras (pitch regions) for each voice—such as soprano, alto, tenor, and bass. However, a statistical analysis showed that Bach was motivated principally by efforts to avoid the crossing of parts. He would rather have had a voice move outside its purported range or retire a voice permanently from the texture than allow the parts to cross.[16] The study also showed that as the number of concurrent parts increases in his music, Bach became ever more vigilant to avoid part-crossing.

Melodic Motion: The Yodel Boundary

Unlike the trill boundary, the yodel boundary plotted in figure 6.3 shows a dramatic downward slope. This means that in order for a sequence of pitches to break apart into two clearly defined streams, tempo is the foremost factor (although interval size still plays a role). If the yodel boundary influences musical organization, then we might expect to see trade-offs between interval size and instantaneous tempo. For example, we might predict that large melodic leaps would be associated with tones of long duration. This idea leads to an interesting research story.

In 1954, psychologist Paul Fitts carried out a simple experiment at the Ohio State University. He presented people with two circular targets on

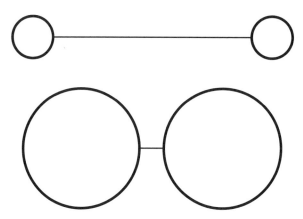

Figure 6.9
Two pairs of circular targets illustrating Fitts's law. Participants were asked to alternate the point of a stylus back and forth as rapidly as possible between two targets. The minimum duration of movement between targets depends on the distance separating the targets as well as target size. The movement is more rapid between the lower pair of targets. Fitts's law applies to all muscle motions, including the motions of the vocal muscles. Musically, the distance separating the targets can be regarded as the pitch distance between two tones, whereas the size of the targets represents pitch accuracy or intonation. Fitts's law predicts that if the intonation remains fixed, vocalists will be unable to execute wide intervals as rapidly as small intervals.

a piece of paper. He then gave them a pen and asked them to move it as rapidly as possible back and forth between the two targets. Fitts varied the distance between the targets as well as the size of the targets. He found that the speed with which people accomplish this task is proportional to the size of the targets and inversely proportional to the distance separating them. The participants achieved faster speeds when the targets were big and close together, as in the lower pair of circles in figure 6.9.

This result seems obvious, if not trivial, but it has interesting repercussions. First, this relationship of speed, distance, and accuracy (target size) turns out to be fundamental. The principle applies to all muscle-powered or autonomous movement and is so basic that scientists refer to it as *Fitts's law*.[17] Because Fitts's law applies to all muscular motion and because vocal production and instrumental performance involve the use of muscles, the law also constrains the generation or production of sound. Imagine, for example, that the circular targets in figure 6.9 are rotated so that they are arranged vertically rather than horizontally. Now consider the targets as representing higher and lower pitches. A small target represents tight accurate tuning; a large target represents sloppy pitch intonation. Now suppose we ask someone to sing back and forth as rapidly as possible between two pitches. Fitts's law tells us that if the intonation (pitch accuracy) remains fixed, vocalists will be unable to execute wide intervals as rapidly as small intervals. It is easier to sing a trill between notes separated by one or two semitones than to sing a "trill" between notes separated by an octave.

There is a parallel phenomenon in vision. In the early 1900s, the German psychologist Adolf Korte carried out a number of studies related to the phenomenon of *apparent motion*. Korte was intrigued by motion pictures, the new technology of his age. How, he asked, can the brain perceive motion by watching a sequence of static pictures? He distilled the problem down to a single pair of lights. Place the lights a certain distance apart and then arrange the electrical circuit so that the lights alternate their on-and-off patterns. As one light is extinguished and the other is illuminated, there will appear to be some movement from one light to the other. Korte discovered that if the lights are too far apart or the speed of switching is too rapid, the effect is lost. Like van Noorden's experiments with alternating pitches, Korte mapped out the effects of different speeds and distances. If the lamps are placed farther apart, then the rate of switching must be slower in order to maintain a sense of apparent motion between the lamps. If the switching rate is too fast or the lamps are placed especially far apart, the viewer sees two independent flickering lights with no sense of intervening motion.[18] Korte summarized his studies as a set of laws regarding apparent motion. It turns out that Korte's third law of apparent motion is equivalent to van Noorden's yodel boundary.[19]

Both van Noorden's trills and Korte's flickering lights are perceptual phenomena. Their research identifies situations where we see or hear either one moving thing or two static things. Korte's third law is a law of how things appear to move (in vision), and van Noorden's yodel boundary is a "law" of how things appear to "move" (in audition). Both phenomena can be related to Fitts's law, which describes how real things (like muscles) move.[20]

In the case of vision, the brain perceives apparent motion only if the visual evidence is consistent with how motion commonly occurs in the real world. If it is implausible for a single real-world object to move along the presumed trajectory, then no apparent motion is perceived. Similarly, an auditory stream is most likely to be perceived when its pitch trajectory conforms to how real (muscle-powered) sound sources behave. If the pitch movement is too rapid—that is, if the pitch movement is faster than what a muscle could achieve—then the listener perceives more than one sound source and the sense of connection between the pitches is lost. In short, it looks like Fitts's law of physical motion provides part of the foundation for the common musical metaphor of "melodic motion."[21]

With this background, we can now return to the idea posed at the beginning of this section. Recall that van Noorden's yodel boundary depends more on speed than on interval size. When the notes alternate rapidly there is a much greater likelihood that they will be perceptually segregated into different streams. If a melody is intended to be heard as a single auditory stream, then large intervals must be treated carefully. Specifically, the music should slow down.

In a cross-cultural sample of several thousand melodies, I found that as the interval size increases, the two tones forming the interval tend to have longer notated durations.[22] This relationship between pitch interval and interval duration is apparent in common melodies such as "My Bonnie Lies over the Ocean" or "Somewhere over the Rainbow": the large intervals at the beginnings of these melodies employ notes that have relatively longer durations. We might call this phenomenon *leap lengthening*. A more recent study by David Temperley at the Eastman School of Music has replicated this observation in string quartets by Haydn and Mozart: the notated music "slows down" for large leaps. Temperley also found that the greatest slowing occurs for the initial note forming the leap.[23] It is as though a long duration tone acts like a launching pad for an ensuing large leap.

Incidentally, the same effect can be observed in performers' use of rubato. Working at the Royal Institute of Technology in Stockholm, Johan Sundberg, Anders Askenfelt, and Lars Frydén found that performers slow down when performing melodic leaps and that listeners prefer the first tone of the leap to be lengthened in duration.[24] In short, when a melody contains a large interval, composers tend to write notes with longer durations and performers add to the effect by slowing down, all consistent with Fitts's law. Perceptually this slowing down increases the likelihood that listeners will hear the notes forming the leap as melodically connected, that is, as forming a single auditory stream.

In light of this discussion, we might now return to the pitch proximity principle and revise it one more time so that it includes the role of van Noorden's yodel boundary:

5. Pitch Proximity Principle (continued)
... *When pitch distances are large, it is possible to prevent stream segregation ("yodeling") by reducing the tempo (i.e., "leap lengthening").*

At least one mystery remains to be explained: How is real yodeling physically possible? If Fitts's law constrains all muscle-powered movement and if movement according to this law results in the perception of a single moving thing, how can the (muscle-powered) human voice produce sounds that lead to the perception of more than one concurrent auditory stream as happens in yodeling?

In order to answer this question, consider a trivial way in which we might seemingly violate Fitts's law. In Fitts's original task, people were asked to move a pen back and forth between two targets. What if we put pens in your left and right hands and position your hands directly over the left and right targets? Even if we move the targets far apart (and make them very small), you can still alternate between the targets very rapidly. The reason why this arrangement doesn't violate Fitts's law is that you are using more than one muscle: you are using two arms rather than just one. Each individual muscle always behaves in accordance with Fitts's law. But it makes sense that independent muscles can produce independent motions, and so convey the impression (either visual or auditory) of more than one action.

With this background, we can now make a prediction about yodeling: *yodeling must involve more than one muscle!* And indeed it does. For adults, two forms of vocal production are common: so-called *modal* (or chest) voice and *falsetto* (or head) voice. In yodeling, singers alternate between modal and falsetto voices. Several muscles are involved in controlling the tightness of the vocal folds. With appropriate coordination, the singer can generate a single sequence of tones that nevertheless segregate into more than one auditory stream. Fitts's law remains unbroken.[25] We hear more than one musical part for the good reason that more than one muscle action is occurring.

In his experiments on pitch-based streaming, van Noorden discovered two boundaries: the trill boundary (beneath which a single stream is always perceived) and the yodel boundary (above which two streams are always perceived). These two boundaries imply that the auditory system makes use of two different heuristics for pitch-based streaming. The yodel boundary is consistent with Fitts's law, suggesting that the auditory system concludes two streams must be present whenever Fitts's law appears to be violated. The trill boundary suggests that the auditory system assumes neighboring pitches must belong to a single stream. Both of these heuristics are fallible, but they are nevertheless serviceable rules of thumb when attempting to assemble a mental picture of the presumed acoustic sources.

Pitch Co-Modulation

The apparent "movement" between successive tones turns out to have repercussions for how minds interpret simultaneous tones as well. As early as 1863, the renowned German scientist Hermann von Helmholtz suggested that similar pitch motion contributes to the perceptual fusion of concurrently sounded tones. Partials that "move" together tend to be grouped together.[26]

Music theorists distinguish various kinds of contrapuntal motions. *Parallel motion* occurs when two parts move in the same direction while maintaining a fixed interval between them. *Similar motion* occurs when two parts move in the same direction but the interval size varies. *Contrary motion* occurs when two parts move in opposite directions, and *oblique motion* occurs when one part moves while the other part retains the same pitch.

The contrapuntal motions described by musicians are meant to describe the behavior of pairs of complex tones. But the same taxonomy of movements can be applied at the level of pairs of partials. An ascending glissando played on a trombone produces dozens of harmonics that all rise together in parallel. Above and beyond the fusion due to the harmonic series, the coordinated ascending frequency movements of the partials further contribute to the perception of a single unified auditory image.

A number of experiments have examined the role of coordinated frequency motion on the formation of auditory images, much of it carried out by Albert Bregman and his collaborators at McGill University.[27] The most exhaustive studies were carried out by Stephen McAdams, one of Bregman's former students.[28] McAdams and other researchers have demonstrated four principles related to the frequency motions of tones. First, parallel motion has a strong tendency to cause the frequencies to fuse into a single auditory stream. Second, this effect is strongest when the tones are harmonically related, as in the case of parallel octave intervals. However, parallel motion increases the tendency to form a single auditory stream, even when the partials are not harmonically related. Gliding pitches that are a major second apart will still exhibit a strong tendency to be heard as a single stream.

Although a much weaker effect, *similar* motions can also contribute to tonal fusion. That is, even when the tones don't move in strictly parallel trajectories, there remains a weak probability that the tones will fuse into a single image provided they are moving in the same direction. Finally, the research has shown that tones moving in the same direction are more likely to fuse into a single stream than when the tones don't move at all.[29] In general, moving pitches are most likely to fuse into a single stream when they are moving strictly in parallel and when the tones are harmonically related.

Conceptually we can distinguish at least three forms of parallel frequency change. *Gliding frequencies* (as in musical glissandi) promote very strong perceptual fusion of all the gliding components. In most music, it is more common to have *discrete frequency shifts*, as when two concurrent tones both abruptly descend by four semitones. Here the tendency for perceptual fusion tends to be weaker than for gliding frequencies. Finally, researchers distinguish *micromodulation* such as occurs with vibrato or with slight pitch instabilities. In this case, all of the partials wobble up and down together. If two instruments play in unison, the two physical sources are more likely to segregate perceptually if one of the instruments plays with vibrato. The coordinated frequency glides will help the partials of the vibrato instrument to pop out. Conversely, if two instrumentalists are able to match their vibratos perfectly, they will increase the unison effect, encouraging listeners to hear the combination as one stream.

When a tone moves up or down in frequency, hearing scientists use the term *modulation*. When two tones move at the same time, we can speak of *co-modulation*. If the tones move in the same direction, the co-modulation is said to be positively correlated, and when they move in opposite directions, the co-modulation is negatively correlated. When the tones move up or down by a fixed interval, the positively correlated co-modulation is precise with respect to log frequency. When the tones move up or down together but the interval size varies, the positively correlated co-modulation is approximate. The terms *semblant* and *nonsemblant* are useful ways of further classifying contrapuntal motions. Both parallel and similar motions are examples of semblant motions, whereas contrary and oblique motions are examples of nonsemblant motions.

In light of the hearing sciences research, we can formulate the following principle:

6. Pitch Co-Modulation Principle
Perceptual union is encouraged when concurrent tones move in a similar pitch direction. Fusion is most enhanced when the movement is precise with respect to log frequency (i.e., a fixed musical interval).

In musical practice, the role of the pitch co-modulation principle depends on the musical goal. If the goal is to avoid the inadvertent fusion of nominally independent musical parts, then a composer might aim to avoid positively correlated pitch motions. Conversely, if the musical aim is to create an "emergent" sound or "virtual instrument" (constructed of many sounds), then a composer might do the reverse, aiming for precisely positively correlated pitch motions.

The most obvious effect of pitch co-modulation on part-writing is evident in the avoidance of parallel fifths and octaves. Musicians won't be surprised to learn that formal studies show that parallel octaves and fifths tend to be avoided in much music-making—notably music written in a polyphonic style where voice independence is highly prized. However, studies of polyphonic music show that the effect of pitch co-modulation can be observed more broadly. First, empirical studies show that parallel octaves are more strongly resisted than parallel fifths. Since octaves promote harmonic fusion more easily than fifths, the compositional practice dovetails beautifully with the perceptual research.

In general, the studies show that nonsemblant motions (oblique and contrary) are favored, while semblant motions are resisted. In addition, the research shows that parallel motion is more strongly resisted than similar motion, and that parallel motion is reduced for all interval sizes, not merely with fifths and octaves. Of course one can point to lots of examples of parallel thirds and sixths in polyphonic works, but surprisingly, these occurrences are less common than would occur in a purely random juxtaposition of musical parts.[30] My original published work on this topic dealt exclusively with the music of J. S. Bach; however, in more recent studies, I have replicated these results with music from a dozen additional polyphonic composers.

Of course, not all musicians aim to avoid perceptual fusion of the musical parts. In many circumstances, composers may wish to encourage instruments to fuse into a sort of virtual instrument. Perhaps the most famous of these instances can be found in a well-known passage in Maurice Ravel's *Bolero* (see figure 6.10). Here Ravel assigns the principal melody to the horn with two piccolos playing in parallel at the intervals of a twelfth and a seventeenth above the horn. That is, the piccolos reinforce the third and fifth harmonics. The celeste also contributes to the parallel harmonic movement. For most listeners, the horns and piccolos meld into a single

Figure 6.10
Passage from Maurice Ravel's *Bolero* (three measures after rehearsal marking "8") showing the use of parallel motion in the piccolos, horn, and celeste. The pitch co-modulation in this orchestration causes an "emergent" or *virtual instrument*.

distinctive timbre. Ravel succeeds in assembling a single complex auditory stream. The effect sounds much like an organ.

In fact, the most common use of pitch co-modulation can be found in organ registration. When an organist selects a single stop, typically one pipe sounds for each key pressed. More commonly, however, organists select several stops at the same time, usually at different octaves (e.g., 16', 8', 4', 2'). Organs also provide mixture stops that combine several pipes tuned as the upper partials in a harmonic series. Mixture stops add brightness to the sound. In electronic music synthesis, this technique of building a composite timbre from independently generated sounds is referred to as *additive synthesis*. When the organist plays an ascending scale, the massive parallel motion between the various stops contributes to the evoking of a single auditory stream, despite the many sounding pipes involved.

Reprise

In previous chapters we considered how static sound images are created. In this chapter, we shifted the discussion to the dynamics of forming images over time. We introduced three further principles that influence the formation of auditory streams: *continuity, pitch proximity*, and *pitch co-modulation.* We learned that discrete sounds are more likely to connect in time when the end of one tone coincides with the beginning of the next. When tones are not contiguous, echoic memory can bridge the gap from one tone to another. If the gap is too long, however, the sense of continuity is weakened or lost.

We learned that both pitch distance and tempo influence whether successive tones tend to be heard as a single stream. We always hear one stream when the amount of pitch change is small (within the trill boundary). We always hear two streams (beyond the yodel boundary) when a pitch/time trajectory violates Fitts's law—a law of how things move. There is also a large intermediate region (between the two boundaries) where listeners may hear either one or two streams depending on the context and the listener's mental disposition. (We'll discuss this further in chapter 13.) We saw a close kinship between how muscles move, the perception of movement in vision, and the sense of connectedness in a musical line.

Finally, we learned that coordinated pitch motion (pitch co-modulation) influences whether concurrent tone sequences combine to form a single stream, or whether they are heard as independent streams.

7 Preference Rules

For musicians trained in the Western musical tradition, the research described in the previous chapters is highly suggestive about the origins of part-writing rules. Rather than relying on an impressionistic sense of the kinship between the perceptual principles and part-writing rules, let's make this relationship explicit. In this chapter, we'll see how most of the conventional core rules of late Baroque part-writing can be inferred from the six principles described in the previous chapters. The purpose of this exercise is to clarify the logic, address the pertinent details, and make it easier to see unanticipated repercussions.

Throughout history, part-writing canons have been expressed in the form of rules. Here we will interpret the research as offering advice rather than as establishing a set of sacred admonitions. When composing music, composers may pursue many goals concurrently, and while some composers on some occasions might embrace the implied goals of voice leading, there are many other possible goals whose pursuit may require some compromises. Accordingly, we will follow in the footsteps of theorists Fred Lerdahl and David Temperley and present "preference rules" rather than "rules."[1]

In the following discussion we will distinguish four kinds of statements: *goals*, *observations*, *corollaries*, and *preference rules*. A *goal* is a desired state; it is an overarching purpose or objective that someone values for some reason. An *observation* is simply an experimentally supported empirical statement, often a general statement summarizing many observations. The observations discussed here arise from the summary principles described in chapters 4 through 6. A *corollary* is a statement that is logically implied by some other statement. If I say, "My shoes are black," a corollary statement might be, "My shoes are not white." A *preference rule* is a statement of advice; it identifies a recommended action or course of conduct—all other things being equal. In this chapter, we are going to recommend various actions that contribute to the achievement of a goal. The empirical observations simply tell us something about the effectiveness of different actions. For convenience, the type of statement is indicated by the abbreviations G (goal), E (empirical observation), C (corollary), and PR (preference rule). In addition each statement is numbered in order to facilitate later discussion (e.g., E5). Finally, we will make a distinction between preference rules that correspond to traditional part-writing rules (PR) and newly inferred rules that are not part of the traditional part-writing canon; new preference rules appear in brackets ([PR]).

The following discussion should not be regarded as some sort of proof of the rules of part-writing. In empirical research there is no such thing as proof. What we informally call "facts" are fallible summaries of fallible observations made by fallible scholars motivated by fallible convictions. All empirical statements are subject to revision, refinement, or retraction. The purpose of this chapter is not to defend or justify traditional voice leading. Rather, it is to clarify the logic and make it easier to see unanticipated repercussions. In short, this exercise should facilitate criticism, not preempt it.

First, I propose a goal for voice leading:

G1. *The goal of voice leading is to facilitate the listener's mental construction of coherent auditory scenes when listening to music. In practical terms, the goal of voice leading is to create two or more concurrent yet perceptually distinct "parts," "voices," or "textures."*

Of course this goal is not some universal musical imperative. It is merely one of an infinite number of potential goals that any musician might wish to pursue (or disregard) as desired. (In chapters 13 and 14 we will see how the preference rules prove helpful in creating musical scenes that extend far beyond the narrow world of chorale-style part-writing; in chapter 16, we will see why this goal might be pleasing for many listeners.)

From our goal, two corollaries follow:

C1. *Effective voice leading requires clear auditory stream segregation between each of the concurrent parts.*

C2. *Effective voice leading requires clear auditory stream integration within each of the individual parts.*

We might begin by recalling the toneness principle. The essence of this principle is that the clearest auditory images evoke unambiguous pitches. We can break this principle into two empirical observations. First:

E1. *Harmonic complex tones evoke clearer auditory images than inharmonic or aperiodic sounds.*

From this we infer a nontraditional preference rule:

[PR1.] **Toneness Rule**. *Prefer the use of harmonic complex tones—tones that evoke clear pitch sensations. Part-writing is less effective using aperiodic sounds or tones with inharmonic partials.*

A few pertinent observations can be made about musical practice relating to this nontraditional rule. First, most of the world's music-making relies on instruments that produce tones exhibiting harmonic (or near-harmonic) spectra. Instruments that produce inharmonic spectra (such as drums) are less likely to be used for voice-leading or melodic purposes. Some inharmonic instruments are manufactured so as to maximize the evoked sensation of pitch. This includes percussion instruments like the glockenspiel, marimba, gamelan, and steel drum. These instruments are more likely to be used melodically because they are able to produce reasonably unambiguously pitched tones. Nevertheless, inharmonic spectra can cause difficulties. Bells produce pitched tones, but their pitches are often ambiguous due to their inharmonic partials. Carilloneurs have long observed that carillons are not well suited to contrapuntal writing.[2]

Continuing with our second empirical observation:

E2. *Tones having the highest toneness evoke pitches centered around 300 Hz, in the region between 80 and 800 Hz.*

From this we infer a rule regarding the preferred compass:

PR2. **Compass Rule**. *The preferred pitch region for part-writing lies between E2 (bottom of the bass staff) and G5 (top of the treble staff), roughly centered near D4.*

Adding now the continuity principle:

E3. *The perception of an auditory stream is facilitated when using contiguous rather than intermittent sound sources. Intermittent sounds are less likely to be perceived as forming a continuous stream when silent gaps exceed about 800 milliseconds.*

This leads to a simple nontraditional preference rule:

[PR3.] **Sustained Sound Rule**. *Prefer parts employing continuous or sustained tones in close succession, with few silences, long gaps, or interruptions.*

The effect of this preference rule in music-making is evident when we compare musical instruments with natural sound-producing objects. Recall from chapter 6 that musical instruments are typically constructed to maximize the period of sustain, and performance practices commonly link successive tones so that the start of one tone is synchronized with the end of the previous tone. Instruments like the banjo, whose sounds decay relatively quickly, tend to play more notes per second than comparable instruments (such as the guitar), whose sounds decay more slowly. The same effect can also be seen in repertoire differences between different instruments. For example, xylophone tones decay faster than marimba tones, and this difference is reflected in the fact that music for xylophone tends to be faster paced than music for marimba.[3]

Adding now the minimum masking principle:

E4. *Auditory masking for any vertical sonority is reduced when the sound energy is spread evenly across the basilar membrane. When typical harmonic complex tones are employed, an even spread of sound energy requires a wider spacing of tones as the sonority descends in register.*

From this we infer one traditional and one nontraditional preference rule for chordal tone spacing:

PR4. **Spacing Rule**. *Prefer wider harmonic intervals between the lower voices in a sonority.*

[PR5.] **Tessitura Rule**. *It is increasingly preferred to have wider intervals separating the lower voices as the sonority becomes lower in overall pitch.*

Recall that research has shown that musical practice is indeed consistent with this latter preference.[4]

Adding now the principle of harmonic fusion:

E5. *Partials are more likely to be assembled into a single auditory image when they conform to a harmonic series. If two complex tones are related by simple integer frequency ratios, their partials are likely to align and so fuse into a single auditory image.*

The degree to which two concurrent tones fuse varies according to the interval separating them:

E6. *Harmonic fusion is greatest with the interval of a unison.*

From this, we infer the traditional injunction against unisons:

PR6. **Unisons Rule**. *Resist shared pitches between voices.*

E7. *Harmonic fusion is second strongest at the interval of an octave, followed by the double octave (fifteenth).*

[PR7.] **Octaves Rule**. *Resist the interval of an octave between two concurrent voices.*

E8. *Harmonic fusion is third strongest at the interval of a perfect twelfth, followed by a perfect fifth.*

[PR8.] **Compound Fifths Rule**. *Resist the interval of a perfect fifth and its compound octave equivalents between two concurrent voices.*

In general:

[PR9.] **Harmonic Fusion Rule**. *Resist unisons more than octaves, octaves more than perfect twelfths, perfect twelfths more than perfect fifths, and perfect fifths more than other intervals.*

Recall from chapter 5 that research has shown that polyphonic practice is consistent with these nontraditional preference rules. Polyphonic composers tended to avoid unisons, but even static octaves and fifths occur less frequently than they would in a random juxtaposition of parts.[5]

Let's turn now to the pitch proximity principle:

E9. *The perception of an auditory stream can be ensured by close pitch proximity between successive tones within a voice.*

The closest possible proximity occurs when there is no pitch movement; therefore:

PR10. **Common Tone Rule**. *If successive sonorities share a common pitch class, prefer to retain this as a single pitch within one voice.*[6]

E10. *The evoking of a single auditory stream is almost certain when pitch movement is within the "trill boundary" (roughly two semitones or fewer).*

Hence:

PR11. **Conjunct Motion Rule**. *If a voice cannot retain the same pitch from one sonority to the next, the preferred motion is by step.*

This preference rule might be restated as a corollary:

C3. **Resist Leaps Rule.** *Resist wide pitch leaps.*

When pitch movements exceed the trill boundary, effective voice leading is best maintained when the yodel boundary is not also exceeded:

E11. *When melodic pitch distances are large, it is possible to prevent stream segregation ("yodeling") by reducing the tempo.*

From this observation we infer a nontraditional preference rule:

[PR12.] **Leap Lengthening Rule.** *When a large leap is unavoidable, long durations are preferred for either one or both of the tones forming the leap.*

Recall that leap lengthening (consistent with Fitts's law) has been observed in a cross-cultural sample of melodies.[7]

Bregman has characterized streaming as a competition between possible alternative organizations.[8] It is not simply the case that two successive pitches need to be relatively close together in order to form a stream. The pitches must be closer than other possible pitch trajectories. That is, previous pitches compete for subsequent pitches:

E12. *When more than one auditory stream is present, subsequent pitches tend to be captured by the nearest existing stream.*

From this observation, we can infer those rules of voice leading that relate to pitch-proximity competition:

PR13. **Proximity Rule.** *Prefer writing parts that move to the nearest chordal tone in the next sonority.*

PR14. **Crossing Rule.** *Resist the crossing of parts with respect to pitch.*

Although other reasons can be suggested for why composers might avoid part-crossing, recall that a study of part-crossing in over one hundred works by J. S. Bach showed that actual musical practice is most consistent with the goal of maintaining close pitch proximity within each part.[9]

PR15. **Overlapping Rule.** *Resist overlapped parts in which a pitch in an ostensibly lower voice is higher than the subsequent pitch in an ostensibly higher voice.*

Once again, this preference rule arises from the pitch proximity principle. Each pitch behaves like a magnet, attracting the nearest subsequent pitch as its successor in the stream.

In light of the pitch proximity principle, pitch leaps would appear to be contrary to good voice leading. However, the idea of pitch competition suggests that some leaps are less disruptive than others. The stream organization will be least disturbed when leaps occur in the highest or lowest voices and when the interval jumps away from the texture:

[PR16.] **Leap Away Rule**. *When large melodic intervals are used, prefer to assign it to the highest or lowest voice, and prefer to leap away from the other voices.*

In chapter 9 we will see evidence that composers do indeed follow this nontraditional preference rule.

Notice that a simple technique for maintaining within-voice pitch proximity and avoiding proximity between voices is to assign each voice or part in a unique pitch region or tessitura. If we restrict voices or parts to their own pitch territories, then part-crossing, overlapping, and other proximity-related problems are reduced. Of course, there are often purely mechanical reasons for restricting an instrument or voice to a given range. But there are also good perceptual reasons for such restrictions, provided the music-perceptual goal is to achieve optimum stream segregation. It is not surprising that in traditional harmony, parts are normally referred to by the names of their corresponding tessituras: the words *soprano, alto, tenor,* and *bass* typically refer to both tessituras and to parts.

Adding now the pitch co-modulation principle:

E13. *Perceptual union is encouraged when concurrent tones move in a similar pitch direction.*

[PR17.] **Semblant Motion Rule**. *Prefer nonsemblant over semblant motion between concurrent parts; that is, resist similar or parallel motions.*

E14. *Fusion is most enhanced when the movement is precise with respect to log frequency (i.e., a fixed musical interval).*

[PR18.] **Parallel Motion Rule**. *If semblant motion is necessary, prefer similar motion over parallel motion.*

Statistical analysis of scores has shown that polyphonic composers in fact write in a manner consistent with a preference for nonsemblant contrapuntal motion. Moreover, efforts to avoid parallel and similar motion are evident for all interval sizes, not just harmonically fused intervals such as unisons, octaves, and perfect fifths. In addition, polyphonic parallel motion is avoided more than similar motion.[10]

Preference Rules from Multiple Principles

Each of the preceding preference rules was inferred from a single perceptual principle. When two or more principles pertain to the same goal, then transgressing several principles will have more onerous consequences than transgressing just one principle. If we violate a single principle, the negative consequences can be minimized if we ensure complete conformity with the remaining principles—but only if all of the principles pertain to the same goal. Since the six perceptual principles all pertain to the goal of forming clear auditory images and streams, we can consider all of their possible combinations. Unfortunately, sixty-four propositions would arise if we formed all the combinations afforded by six principles. Rather than consider all sixty-four, we will consider only a handful.

If we link the harmonic fusion and pitch proximity principles, we have:

E5&E9. *The detrimental effect of harmonic fusion can be reduced by ensuring close pitch proximity within the parts.*

This leads to two nontraditional preference rules:

[PR19.] **Oblique Preparation Rule**. *When approaching unisons, octaves, fifteenths, twelfths, or fifths, it is preferable to retain the same pitch in one of the voices (i.e., approach by oblique motion).*

[PR20.] **Conjunct Preparation Rule**. *If it is not possible to approach unisons, octaves, fifteenths, twelfths, or fifths by retaining the same pitch (oblique motion), then step motion is the next preferred approach.*

Linking the harmonic fusion and pitch co-modulation principles yields:

E5&E13. *Harmonic fusion is further encouraged when harmonically fused intervals move in a similar direction.*

From this we may infer the rule:

[PR21.] **Nonsemblant Preparation Rule**. *Resist similar pitch motion in which the voices employ unisons, octaves, or perfect twelfths/fifths (e.g., when both parts ascend beginning an octave apart, and end a fifth apart.)*

E5&E13. *Harmonic fusion is most encouraged when harmonically fused intervals move in a parallel pitch direction.*

From this observation, we infer the well-known prohibition against parallel fifths and octaves:

PR22. **Perfect Parallels Rule**. *Resist parallel unisons, octaves, twelfths, and fifths.*

Joining the harmonic fusion, pitch proximity, and pitch co-modulation principles, we find that:

E5&E9&E13. *When moving in a similar direction toward a harmonically fused interval, the detrimental effect of harmonic fusion on stream organization can be alleviated by ensuring proximate pitch motion.*

From this observation, we can infer a basic form of the traditional injunction against direct or hidden intervals:

PR23. **Direct Intervals Rule**. *When approaching unisons, octaves, twelfths, fifths, or fifteenths by similar motion, at least one of the voices should preferably move by step.*

In textbooks on part-writing, most theorists apply the direct intervals rule to both fifths (twelfths) and octaves (fifteenths). A minority of theorists restrict this injunction to approaching just octaves (fifteenths)—that is, "direct octaves." Since octaves lead to harmonic fusion more easily than perfect fifths, the perceptual principles account for why there would be greater unanimity of musical opinion prohibiting direct octaves compared with prohibiting direct fifths.

Notice that PR23 differs from conventional statements in two respects. Most theorists restrict the direct intervals rule to part-writing involving just the bass and soprano voices. In addition, most theorists specify that step motion should occur in the soprano rather than in the bass. In chapter 12 we will revisit the direct intervals rule. We'll see that perceptual tests of the rule push us beyond the confines of Baroque practice and lead to a broader way of thinking about musical texture. This will have repercussions for considering jazz and pop arranging, large-scale orchestral textures, electroacoustic music, and sonic design.

Reprise

In this chapter we used six perceptual principles to infer twenty-three preference rules. Ten of the rules are conventionally part of the core part-writing canon dating from the late Baroque period. What about the thirteen novel preference rules? Empirical research has already established that polyphonic composers write in a manner consistent with all of these additional rules. In short, there appears to be a very close relationship between research on the formation of auditory streams and part-writing practices inherited from the seventeenth century and earlier.

8 Types of Part-Writing

We are not quite finished discussing the core of voice-leading rules. The six perceptual principles described in the previous chapters are not the only ones that influence the formation of auditory streams. In this chapter we consider four additional principles; however, these other principles don't show a consistent influence on musical organization. Musicians appear to treat these other principles as compositional "options." By the end of this chapter, we will see that these auxiliary principles shape music-making in perceptually distinctive ways and so help to distinguish some basic musical genres.

Onset Synchrony

Raise your left foot off the ground and prepare to strike the floor. Now hit the floor with your foot and slap your right hand against your chest at the same time. Physically, you will be creating two different sounds. A person watching you might conclude that you are a bit odd, but they would be in little doubt that you generated two concurrent sounds. However, anyone not watching you would have likely concluded that there was just a single sound event. Sounds whose onsets are coordinated in time are much more likely to be interpreted by the auditory system as constituents of a single complex sound rather than two distinct sound images.

Both the synchronized starting and stopping of partials contribute to the formation of a single auditory image. However, starting at the same time is more important than ending at the same time. When sounds have uncoordinated onsets, they are likely to be perceived as distinct or separate events.[1]

The concept of synchronization raises the question of what is meant by "at the same time." Under ideal listening conditions, people can distinguish two clicks separated by as little as 20 milliseconds.[2] However, in more realistic listening situations, onset differences can be substantially greater and yet retain the impression of a single onset. Even for natural percussive sounds, the onsets of the various partials may be spread over 80 milliseconds or more.[3] For less percussive sounds, it may take much longer for all of the partials to appear. In the case of echoes, more than 100 milliseconds between successive sounds may be required in order for the echoes to be perceived as distinct events.[4]

The Dutch musicologist Rudolf Rasch has collected a wealth of data on the synchronization of note onsets in music.[5] With a small ensemble such as a string quartet, even the most precisely coordinated onsets are typically spread over a range of 30 to 50 milliseconds. In experiments with quasi-simultaneous musical tones, Rasch found that the effect of increasing asynchrony is initially to add transparency to multitone stimuli rather than to evoke separate sound images.[6] Onset asynchronies have to be comparatively large before separate sound events are perceived. In general, sounds whose onsets start within a tenth of second are more likely to be grouped together and heard as a single sound.

Drawing on this research, we might formulate the following principle:

7. Onset Asynchrony Principle

If a composer intends to write music in which the parts have a high degree of perceptual independence, then synchronous note onsets ought to be avoided. Onsets of intentionally distinct sounds should be separated by 100 milliseconds or more.

Composers have little control over the asynchronies arising from variations due to performance. But they can introduce intentional asynchronies by notating them in a musical score. At a tempo of eighty quarter notes per minute, a sixteenth note is roughly 190 milliseconds in duration. If the notation instructs two instrumentalists to play tones whose onsets are separated by a sixteenth duration, this is plenty of time for listeners to hear the onsets as distinct events. At the same tempo, a thirty-second note is less than 100 milliseconds in duration. If the musical notation instructs two instrumentalists to play tones separated by less than a thirty-second note, then the probability increases that listeners will hear a single sound onset. In fact, music rarely notates asynchronies of a thirty-second note or shorter. Although musical notation provides only a crude indication of the timing of real performances, the research suggests that synchronously and asynchronously notated events are likely to be perceived in the appropriate ways.

Several studies have examined musical notation for evidence related to the onset synchrony principle. These studies have shown that musical practice is equivocal: some music conforms to this principle, but much music does not. The music of the sixteenth-century German composer Michael Praetorius provides an example of music that conforms to the principle. Rasch carried out an analysis of vocal works by Praetorius and measured the ratio of synchronous to asynchronous onsets in the notation. He found that the onset synchrony between the parts decreases as the number of voices increases.[7] With increasing numbers of parts, the potential for stream confusion increases, so it makes sense that a composer would decrease the proportion of synchronous onsets in order to help listeners segregate the individual voices in the musical texture.

You might suspect that as the number of parts increases, the proportion of asynchronous onsets would naturally increase. So how do we know that polyphonic composers really make an effort to minimize onset synchrony? In order to demonstrate this, we need to compare the actual music with some sort of "control" music—music that might have been written instead. A useful technique for generating such control music is the autophase method. Suppose we have a two-part musical work notated on a single long strip of paper. We take the ends of the paper and tape them together to form a large paper loop. Now we take a pair of scissors and cut the two musical lines apart, which leaves us with two circular loops of paper. We can rotate these two loops with respect to each other like a screw-top lid on a jar. Initially the two loops are synchronized as in the original score. We might then shift the upper part ahead one measure. This means that the upper voice will be playing measure 1 at the same time that the lower voice is playing measure 2. Of course, except for the original position (0 degrees), all other rotations between the parts produce musical nonsense. But notice that this nonsense preserves a number of musical properties: the melodic intervals in both parts are perfectly preserved, and both parts maintain the original rhythm, the same meter, the same phrasing and note durations, and so on. The only thing that is destroyed by rotating the parts is the relationship between the parts—and it is precisely the relationship that we are interested in. For each point of rotation, we have created a new piece of "control music." For each control piece, we can tally up the proportion of synchronous to asynchronous onsets between the parts and compare these rotated values with the amount of synchronization when the parts are at zero degrees (that is, as originally written).

Figure 8.1 shows onset synchrony autophases for a random selection of ten of Bach's two-part keyboard Inventions (BWVs 772–786). In addition, an average of all ten onset synchrony functions is plotted (solid line). The values in the horizontal center of the graph (phase 0) indicate the degree of onset synchrony for each work as originally written by Bach. All other values show the degree of onset synchrony for control pieces in which the parts have been shifted with respect to each other, measure by measure. Notice a tendency for all of the lines to dip at zero degrees. This means that the fewest synchronous onsets between the parts occur as Bach wrote it. Other relationships typically result in a greater proportion of synchronous note onsets between the parts. This graph doesn't prove that Bach intended to minimize the degree of onset synchrony in his polyphonic works, but the results are consistent with this interpretation.[8]

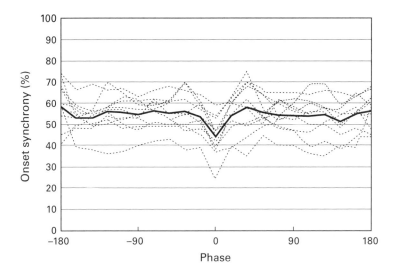

Figure 8.1
Onset synchrony autophase functions for a random selection of ten of Bach's two-part keyboard Inventions (BWVs 772–786) (Huron, 1993a). Values plotted at zero degrees indicate the proportion of onset synchrony for the actual works. All other phase values indicate the proportion of onset synchrony for rearranged music, controlling for duration, rhythmic order, and meter, etc. The dips at zero degrees are consistent with the hypothesis that Bach avoided synchronous note onsets between the parts. The solid line plots an average onset synchrony function for all ten works.

Types of Part-Writing

Of course, not all music is written to reduce synchronous onsets. In non-polyphonic works, composers routinely contradict this principle. Figure 8.2 compares onset autophases for two contrasting works. The first graph shows an autophase for Bach's three-part Sinfonia no. 1 (BWV 787). The second graph shows an autophase for the four-part hymn "Adeste Fideles" ("Oh Come All Ye Faithful"). When we rotate two parts with respect to each other, we get a two-dimensional graph as in figure 8.1. With three-part music, there are now three parts to rotate with respect to each other, and this results in a three-dimensional autophase graph. In the 3D graph, the vertical axis displays the proportion of synchronous onsets for the various conditions. The actual amount of onset synchrony in Bach's Sinfonia is plotted at (0,0) at the front of the graph. All other values show the degree of onset synchrony for control conditions in which the parts have been systematically shifted with respect to each other, measure by measure. The four-part "Adeste Fideles" produces a four-dimensional autophase graph; it is impossible to display on the page, so the graph has been simplified to three dimensions by phase-shifting only two of the three voices with respect to a stationary soprano voice. In order to simplify the graph, the actual onset synchrony (0,0) is plotted at the back of the graph (it appears as a tall spike).

As you can see, in Sinfonia no. 1 there is a significant dip at (0,0) consistent with the notion that Bach was actively avoiding onset synchrony. However, in "Adeste Fideles," there is a significant peak at (0,0) (at the back of the graph). This result implies the opposite motivation, suggesting that synchronous onsets were actively sought. The two graphs demonstrate that any realignment of the parts in Bach's Sinfonia no. 1 would result in greater onset synchrony, whereas any realignment of the parts in "Adeste Fideles" would result in less onset synchrony. Why are these two pieces so different?

Musicians recognize that these two works differ with regard to musical texture. Bach's Sinfonia No. 1 is *polyphonic* in texture, whereas "Adeste Fideles" is *homophonic* in texture. The distinction between polyphony and homophony seems so natural that we musicians can easily take the distinction for granted. So just what is the difference between these two classic textures?

In 1989 I carried out a statistical study designed to identify the main differences among four classic musical textures. My study employed over a hundred works by several composers whose textures were independently

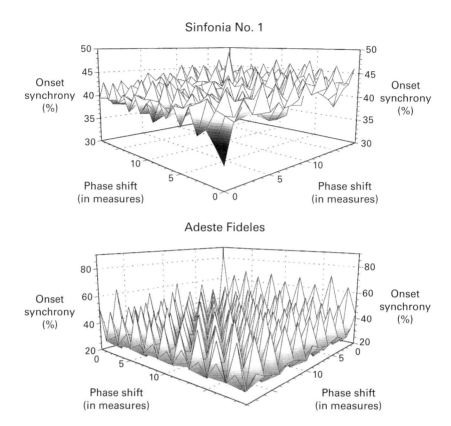

Figure 8.2
Three-dimensional onset synchrony autophase graphs comparing polyphonic and homophonic works: (a) three-part Sinfonia no. 1 (BWV 787) by J. S. Bach; (b) four-part hymn "Adeste Fideles." The vertical axis indicates the onset synchrony (measured according to method 3 described in Huron, 1989a). The horizontal axes indicate measure-by-measure shifts of two of the voices with respect to the remaining voice(s). Figure 8.2a shows a marked dip at the origin (front), whereas figure 8.2b shows a marked peak at the origin (back). The horizontal axes in the two graphs have been reversed to preserve visual clarity. The graphs formally demonstrate that any realignment of the parts in Bach's Sinfonia no. 1 would result in greater onset synchrony, whereas any realignment of the parts in "Adeste Fideles" would result in less onset synchrony.

classified by musicians as *monophonic, homophonic, polyphonic,* or *heterophonic*. A work is considered monophonic when it consists of a single unaccompanied line of sound (as in the unison singing of a folksong). A work is regarded as polyphonic when it involves several concurrent independent lines (as in an organ fugue). A homophonic work consists of successive block sonorities (as in a sequence of strummed guitar chords). A heterophonic work involves several instruments or voices performing simultaneous variations of the same musical line. (Heterophonic works are extremely rare in Western music; the effect might be described as a braid of interleaved renditions of a single melody.)

In my study, each work was characterized according to six criteria. For example, one of the criteria was the amount of parallel and similar contrapuntal motion.[9] I used a technique known as discriminant analysis in order to determine which factors best distinguish the various types of textures.[10] Of the factors studied, I found that by far the most important discriminator between homophonic and polyphonic music is the degree of onset synchrony. The most important difference between polyphony and homophony is exactly the difference we have seen between Bach's Sinfonia no. 1 and "Adeste Fideles."

Why Does Homophony Exist?

The existence of homophonic music raises a paradoxical perceptual puzzle. On the one hand, much (though not all) homophonic music obeys the traditional part-writing rules outlined in chapter 2. Baroque-style chorales, for example, are prototypical homophonic textures whose individual voices conform to conventional part-writing rules. However, if the part-writing is intended to encourage perceptually independent voices, why wouldn't the parts also be composed so they maintain independent rhythms? That is, why wouldn't composers make an effort to minimize synchronous onsets between the parts? Notice that this question is tantamount to asking the question: Why isn't all multipart music polyphonic in texture?

First, it is appropriate to make a distinction between two types of homophonic textures. Many homophonic works, like Baroque-style chorales, do indeed conform to conventional part-writing rules while exhibiting high onset synchrony. Yet a second homophonic practice ignores or reverses the traditional part-writing considerations. These include homophonic passages that feature massive amounts of parallel motion, or ignore pitch proximity and other principles. A sequence of strummed guitar chords may be produced without any concern for how the "voices" stream between chords; even the number of tones in successive chords is likely to vary implying that "voicing" is not a concern.

The first form of homophony is akin to polyphony except for the disregard of the onset synchrony principle. The second form of homophony is akin to monophony except that successive sonorities consist of multiple pitches (i.e., chords) instead of single pitches. For convenience, we might refer to the first as "polyphonic-like homophony" and the second as "monophonic-like homophony." We will consider monophonic-like homophonic textures in greater detail in chapter 13. For the current discussion, it's appropriate to consider some general points.

When we pursue some goal, we sometimes compromise the pursuit of it in order to pursue another goal at the same time. When visiting a restaurant for example, choosing an item from the menu is not simply a matter of selecting what one believes to be the most delicious offering. You might also take into account the cost of the item, the nutritional content, or the novelty value, for example. When composers create a musical work, they often have more than one goal in mind. In the case of homophonic music, musicians clearly want the various parts to have (mostly) synchronous onsets. There are probably good reasons for this.

The preference for synchronous onsets could originate in any of a number of music-related goals, including social, historical, stylistic, aesthetic, formal, perceptual, emotional, pedagogical, or other goals that the composer may be pursuing. In trying to account for homophonic music, it is reasonable to assume the existence of some additional goal that has a higher priority than the improvement of stream segregation that would result from asynchronous onsets. More precisely, this hypothetical goal must have repercussions for the *temporal* organization of the music. Two plausible goals come to mind.

One of the differences between Sinfonia no. 1 and "Adeste Fideles" is that one is instrumental and the other is vocal. A perennial concern in vocal music is the intelligibility of the words. In multipart music, vocal texts are better understood when all voices utter the same syllables concurrently, so we might predict that vocal music is less likely to be polyphonic than purely instrumental music. Moreover, when vocal music is truly polyphonic, it may be better to use melismatic writing and reserve syllable onsets for occasional moments of coincident onsets between several parts. In short, synchronous onsets might be expected to be favored in vocal music, simply as a way of promoting the intelligibility of the text. Polyphonic-like homophony might therefore arise as a compromise enabling the pursuit of two concurrent perceptual goals: the perceptual independence of the concurrent voices and the preservation of the intelligibility of the text.

But there is at least one other possibility. In ways not yet understood, music with a strong sense of rhythm can evoke pleasure. Of course, virtually all polyphonic music is composed within a metric context. But the rhythmic effect of polyphony is typically that of an intricate braid of rhythmic figures. The rhythmic energy associated with marches and dances (such as pavans, gigues, and minuets) cannot be achieved without some sort of rhythmic uniformity. Bach's three-part Sinfonia no. 1 appears to have little of the rhythmic drive evident in "Adeste Fideles." In short, synchronous onsets might be expected to be favored in music, simply as a way of building rhythmic momentum and energy. Polyphonic-like homophony might therefore arise as a compromise enabling the pursuit of two concurrent perceptual goals: the perceptual independence of the concurrent voices, and creating rhythmic energy. As in the case of the text-setting proposal, this conjecture generates a prediction: that works (or musical forms) that are more rhythmic in character (e.g., cakewalks, sambas, waltzes) will be less polyphonic than works following less rhythmic forms.

If either of these accounts holds merit, then it suggests that genres like Baroque-style chorales represent a mix of compositional goals. The structure of the music reflects the composer's aims.

Yet More Preference Rules

Having discussed possible origins for homophony, let's return to the polyphonic case. If a composer's goal is to maximize the perceptual independence of the concurrent voices, then minimizing synchronous onsets clearly is appropriate. Restating the empirical observation:

E15. *Concurrent tones tend to fuse into a single auditory image when their onsets are synchronous.*

From this we can infer the following nontraditional preference rule:

[PR24]. **Asynchronous Onsets Rule.** *Prefer asynchronous onsets for concurrent voices.*

Notice that some synchronous onsets will threaten perceptual independence more than others. As we saw in the previous chapter, several concurrent voice-leading transgressions are likely to cause the biggest problems, so it is important to consider how the onset synchrony principle might combine with the other core principles. Once again, space limitations prevent us from considering all of the interactions. Instead, let's consider what happens when we combine the onset synchrony principle with the principle of harmonic fusion. Perceptual experiments carried out by Dutch musicologist Joos Vos have shown an interaction between onset synchrony and harmonic fusion in the formation of independent auditory images. Asynchronous onsets contribute more to forming independent images in the case of perfect intervals (intervals that are more susceptible to harmonic fusion).[11] Vos's work establishes that:

E5&E15. *In the approach to a harmonically fused interval, the potential for harmonic fusion can be reduced by using asynchronous tone onsets.*

From this we can infer the following nontraditional preference rule:

[PR25]. **Asynchronous Preparation Rule.** *When approaching unisons, octaves, twelfths, or fifths, prefer asynchronous note onsets.*

The most problematic onsets are those involving intervals already disposed toward harmonic fusion, a problem illustrated in figure 8.3. Passage A involves successive thirds; since these intervals show little tendency to promote harmonic fusion, the *synchronous* onsets raise no concern. However, passage B approaches a perfect fifth with synchronous onsets. The research suggests that the voice leading would be improved by an asynchronous preparation of perfect intervals as in passage C, which rearranges the progression so that the perfect fifth is "prepared." So does music exhibit this tendency? Do composers follow this preference rule?

Yes. In a study of harmonic intervals in polyphonic works, I looked at which intervals are approached synchronously and which are approached asynchronously. Consistent with traditional theory, dissonant intervals (like seconds, sevenths, and ninths) tend to be asynchronously "prepared."

Figure 8.3
Three passages illustrating (a) synchronous movement to a nonfused interval, (b) synchronous movement to a harmonically fused interval, and (c) asynchronous preparation of a tonally fused interval.

That is, most dissonant intervals occur when first one pitch appears, and then after a delay, the second pitch forming the interval appears. By contrast, most imperfect consonances (e.g., thirds, sixths, and tenths) have synchronous onsets: both notes tend to sound simultaneously. However, most perfect consonances (e.g., fourths, fifths, and octaves) are approached asynchronously. Like the dissonant intervals, intervals most prone to tonal fusion tend to be "prepared."[12]

There is more to the story. In a separate study, I examined the approach to various harmonic intervals in two different repertoires by J. S. Bach—polyphonic works (fugues) and more homophonic works (chorales). In both the fugues and the chorales, dissonant intervals tend to be prepared using asynchronous onsets. Similarly, in both repertoires, most imperfect intervals occur with synchronous note onsets. However, for the perfect intervals, the two repertoires produced contrasting results. Where the majority of perfect intervals in the polyphonic music are approached asynchronously, the majority of perfect intervals in the more homophonic music are approached synchronously. This interaction between onset synchrony and harmonic fusion for the two different repertoires is illustrated in figure 8.4.

What figure 8.4 doesn't show is that for the perfect intervals alone, there is a positive correlation between the degree to which an interval promotes harmonic fusion and the likelihood of asynchronous preparation. Unisons and octaves are more likely to be approached asynchronously than perfect twelfths and fifths, followed by elevenths and fourths.[13] In short, J. S. Bach composed in a manner consistent with the nontraditional preference rule PR25, but only in his more polyphonic works.

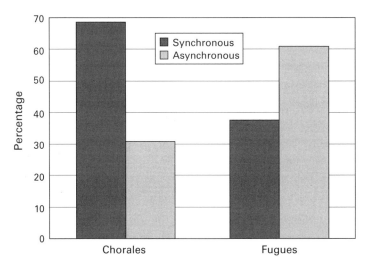

Figure 8.4
Comparison of harmonically fused intervals in chorales and fugues by J. S. Bach. Only perfect harmonic intervals are plotted (e.g., unisons, fourths, fifths, octaves, twelfths). In the more homophonic chorale repertoire, most perfect intervals are formed synchronously; that is, both notes tend to begin sounding at the same moment. But in polyphonic textures, most perfect intervals are approached with asynchronous onsets (one note sounding before the onset of the second note of the interval). The two repertoires show no differences when approaching imperfect and dissonant intervals.

Scene Density

Listeners do not have an unlimited capacity to track multiple concurrent lines of sound. In a perceptual experiment, I asked expert musicians (performers and teachers) to listen to recordings of polyphonic music and indicate how many concurrent parts they heard. On a keypad, there were numbered keys from 0 to 10. When they heard silence, they were to press and hold the key marked zero. When they heard only a single musical part, they were to press and hold the key marked one—and so on. The music was polyphonic organ music containing up to five concurrent parts.

One of my findings was that as the number of voices increases, expert listeners are slower to recognize the addition of new voices. For example, when only a single part is sounding, musicians are very fast to move from key 1 to key 2 when a second part enters. But when the texture increases from three to four parts, musicians are noticeably slower responding with the appropriate keys. I also found that as the number of voices increases, musician listeners are more likely to make mistakes in identifying the correct number of voices present. Figure 8.5 plots the summary results.[14] The black columns plot the average estimation errors for different numbers of concurrent parts. The gray columns indicate the average error rates for recognizing entries of single new voices. For musical textures in which all the parts have similar timbres, the accuracy of identifying the number of concurrent voices drops dramatically at the point where a three-voice texture increases to four voices. Beyond three voices, tracking confusions are commonplace. In a subsequent study, Richard Parncutt at the University of Graz in Austria asked listeners simply to count the number of tones in various chords. Even when octave doubling is avoided, Parncutt found that listeners make significant errors once the number of simultaneous tones exceeds three.[15]

In both the Huron and Parncutt experiments, the individual parts or voices had similar or identical timbres. So what happens when a range of different timbres is used? Michael Schoeffler and his colleagues at the International Audio Laboratories in Erlangen, Germany carried out a test of 1,230 listeners (both amateur musicians and non-musicians). In their experiment, listeners were instructed simply to identify the number of instruments present in various recorded excerpts. For example, one of their two-instrument passages consisted of a duet for cello and bassoon. One of their six-instrument passages included violin, oboe, clarinet, French horn, bassoon, and contrabass. Surprisingly, they found the same results when listeners counted the number of instruments. In general, amateur musicians did better than nonmusicians. But once the musical texture exceeded three instruments, all listeners had grave difficulties tracking the individual lines. In all three studies, there was a tendency to underestimate the number of parts, voices, or instruments present.[16]

The difficulties listeners have in tracking more than three concurrent voices may be symptomatic of a broader perceptual limitation. As it turns out, similar difficulties are evident in vision. A variety of counting and estimation tasks show a marked increase in confusion when the number of visual items exceeds three.[17] The effect can be seen even in infants less than a year old. Infants can easily discriminate between two and three visual items, but their performance degrades when discriminating three from four items and reaches chance performance when discriminating four from five

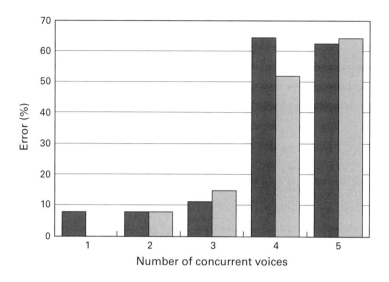

Figure 8.5
Voice-tracking errors while listening to polyphonic music. Black columns: average estimation errors for textural density (number of polyphonic voices). Gray columns: unrecognized single-voice entries in polyphonic listening. The data show that tracking confusions for listeners to polyphonic textures employing relatively homogeneous timbres are common when more than three voices are present. *Source:* Huron (1989b).

items.[18] Since babies haven't yet learned to count, this work suggests that the perceptual confusions evident when more than three visual or auditory images are present arise from a low-level perceptual constraint that is not mediated by cognitive skills such as counting. In the 1920s, the Swiss-French psychiatric researcher Alice Descoeudres had already observed this effect, and referred to it as the *un, deux, trois, beaucoup* phenomenon.[19]

The perceptual experiments simply reinforce observations that have been made informally by musicians for many decades. A number of musicians, including composer Paul Hindemith, have suggested that even the most talented musicians cannot clearly follow more than three simultaneous lines of polyphony.[20]

Drawing on this research, we formulate the following principle:

8. Limited Density Principle
It is difficult for listeners to distinguish more than three concurrent voices or parts.

Of course most harmony writing consists of four or more parts, and most harmony textbooks focus exclusively on four-part writing. A large proportion of sixteenth- and seventeenth-century polyphonic music is in five parts. So once again, the question that arises regarding this principle is why it seems to be rarely followed.

A good place to begin is to investigate the actual number of concurrent voices for different musical works. What do composers actually do? Consider once again the polyphonic works of J. S. Bach. Figure 8.6 reproduces the main result from my study of textural density in Bach's contrapuntal practice.[21] In order to interpret this figure, we must first distinguish the nominal number of parts in a work from the actual number of parts. For example, a fugue might nominally be "in four parts." But this doesn't mean that four parts are present from the beginning to the end of the work. We are interested in the typical or average textural density, not necessarily the maximum density.

We must also be careful about assuming that a single notated "part" is the same as an auditory stream. Chapter 6 discussed how a single sequence of tones can break apart into more than one auditory stream. Since Bach was fond of pseudo-polyphony, we cannot assume that a single notated part will evoke a single auditory stream. Figure 8.6 takes into account all of these possibilities. The figure plots the estimated number of auditory streams versus the nominal number of voices or parts for a large body of polyphonic music by Bach.[22] If an N-part work truly maintains a mean textural density of N auditory streams, then the plotted value for the work would be expected to lie near the diagonal dotted line. In order for a value to lie above the dotted line, the work will typically contain parts that exhibit pseudo-polyphonic activity and avoid idle parts (rests). Conversely, in order for a value to lie below the dotted line, the work will typically contain significant periods of rest and will refrain from pseudo-polyphonic writing.

Figure 8.6 shows that as the number of nominal voices increased, Bach gradually changed his compositional strategy. For works employing just one or two parts, he endeavored to keep the parts active (few rests, of short duration) and boost the textural density through pseudo-polyphonic writing. For works having four or more nominal voices, Bach reversed this strategy. He avoided writing pseudo-polyphonic lines and retired voices from the texture for longer periods of time. Figure 8.6 reveals a change of strategy consistent with a preferred textural density of three auditory streams.

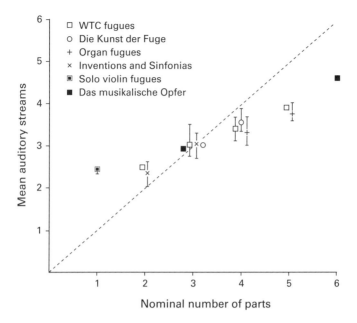

Figure 8.6
Relationship between nominal number of voices or parts and mean number of estimated auditory streams in 108 polyphonic works by J. S. Bach. Plotted points indicate average values for each repertoire; vertical bars indicate data ranges. The dotted line shows a relation of unity slope, where an N-voice polyphonic work maintains an average of N auditory streams. The graph shows evidence consistent with a preferred textural density of three auditory streams. *Note:* Average stream density for each work was calculated according to a method described and tested experimentally in Huron (1989a, chap. 14).

While Bach regularly ventured into textures containing four or more concurrent parts, his typical polyphonic texture is three concurrent voices. At least in the case of J. S. Bach, the music-making seems consistent with the perceptual research: it appears that Bach tended to maximize the number of auditory streams while simultaneously minimizing the number of occasions when the textural density would exceed the listener's ability to track the concurrent parts.

But what about other composers and other repertoires by Bach? For example, Bach's chorale harmonizations are nearly all written in four parts from beginning to end. And the majority of works by other composers maintain more than three concurrent parts.

Types of Part-Writing

Once again, we might entertain the possibility that there is another goal that takes precedence over the goal of the perceptual independence of the musical parts. What goal might favor writing in four or more parts despite the difficulties this creates for listeners in tracking the individual voices or parts? A longstanding enthusiasm in Western music-making has been the phenomenon of triadic harmony. If a composer wants to produce a clear harmonic progression, then it is desirable to have the root, third, and fifth present in each successive chord. Regularly omitting one of the chord members (like the third or the root) tends to weaken the perceived chord or harmonic function.

In general, if a composer wants to employ chords using all triad members and also wants to connect the parts by following the conventional rules of part-writing, then a three-voice texture will prove unduly restrictive. This lack of freedom is made worse if the composer also wants to employ a homophonic texture in which synchronous onsets and block chords prevail. Expanding to a four- or even five-voice texture may be a compromise that allows the composer to pursue both harmony and voicing leading goals at the same time.

What evidence do we have that harmony might sometimes trump voice leading as an overarching goal? A telling example can be found in the so-called *frustrated leading tone* (as illustrated in figure 8.7).[23] This is a not uncommon harmonization where the leading tone in the tenor drops down to the dominant pitch rather than rising up to the tonic. There are two good reasons why the tenor pitch should rise upward to the tonic. The first is the pitch proximity principle; the second is the resolving tendency

Figure 8.7
Illustration of the "frustrated leading tone." In the key of C, the leading tone (B) does not resolve upward to the tonic, but instead drops down to the dominant in order to allow the tonic chord to be spelled with all three chord factors (root, third, and fifth). This not uncommon harmonization suggests that in many circumstances, musicians deem full triadic harmony to be a more important goal than optimum voice leading.

of the leading tone (which will be discussed in chapter 10 on voice leading). Conversely, there are two reasons why the tenor might drop down as illustrated. One is that tripling the pitch C increases the likelihood of harmonic fusion. However, most musicians would claim that the main reason why a frustrated leading tone harmonization is acceptable is that it completes the tonic triad so that all chord members (root, third, fifth) are present. The goal of optimum voice leading is compromised (but not abandoned) in order to achieve a goal that is deemed more important.

If a composer wishes to create a polyphonic-like homophonic texture in which the chords are complete, then four parts is the minimum texture that makes this feasible.[24] Notice, incidentally, that a polyphonic texture makes three-part harmonic writing much more tractable. Since synchronous onsets are avoided in polyphonic music, harmonies are more commonly implied rather than explicitly stated through the use of block triads.

Timbre

Intuitively, we would expect different tone colors to influence how listeners hear auditory streams.[25] If a violin and clarinet play at the same time, listeners will tend to connect successive violin tones into one stream and successive clarinet tones into a different stream. The influence of timbre on the formation of auditory streams is nicely illustrated in a demonstration devised by David Wessel at the Institut de Recherche et Coordination Acoustique/Musique in Paris.[26] Wessel created a repeated sequence of three rising pitches (see figure 8.8). Successive tones were generated so that every second tone was synthesized using one of two contrasting timbres. (The different timbres are illustrated using solid and open noteheads.) Notice that although the pitch sequence consists of three ascending tones, each of the timbre sequences produces a descending tone sequence.

Figure 8.8
Schematic illustration of Wessel's (1979) illusion. A sequence of three rising pitches is constructed using two contrasting timbres (marked with open and closed noteheads). As the tempo of presentation increases, the perception of an ascending pitch figure is replaced by two independent streams of descending pitch, each distinguished by a unique timbre.

If timbre-based streaming takes precedence over pitch-based streaming, then listeners ought to hear two distinct descending pitch sequences rather than a single ascending pitch sequence. Wessel found that as the speed of repetition increases, there is a greater tendency for listeners to hear descending pitch sequences. Also, Wessel found that the effect is most pronounced when the two timbres used are most contrasting. The tendency for different timbres to form separate streams is related to listener judgments of similarity. Listeners tend to segregate sonic events whose tone colors are judged most dissimilar; conversely, listeners tend to link together sonic events whose timbres are judged most alike.[27]

Drawing on this research, we might formulate the following principle:

9. Timbral Differentiation Principle

Differences of timbre contribute to the perceptual independence of concurrent voices or parts.

Again, the question that immediately arises regarding this principle is, Why do composers routinely ignore it? Polyphonic vocal and keyboard works, string quartets, and brass ensembles maintain remarkably homogeneous timbres. Of course, some types of music do seem to be consistent with this principle.[28] The woodwind quintet is a common genre based on heterogeneous instrumentation, and many modern works for small ensembles call for an eclectic mix of instrumental timbres. Nevertheless, tonal composers have not commonly assigned different timbres to each of the various voices. A number of reasons might be cited for this practice. Common keyboard instruments are limited in their capacity to assign different timbres to the different parts. Dual-manual harpsichords and pipe organs do allow some timbral differentiation. In the case of organ music, for example, a common Baroque genre was the *trio sonata* in which two treble voices are assigned to independent manuals and a third bass voice is assigned to the pedal division. Organists use different stops for each of the three polyphonic voices. The dual manual harpsichord can also provide contrasting registrations, although only two voices can be distinguished in this manner. The piano, however, provides little opportunity to distinguish voices by different timbres. Like the piano, vocal ensembles also have a limited capacity to produce different timbres for each musical part.

Another reason composers might favor homogeneous timbres arises from practical difficulties recruiting a specified assortment of instrumentalists.

Suppose that a composer creates a duet specifically for flute and oboe. The flute tends to be very quiet in the low register but brighter in the middle register. The oboe, by contrast, can be quite loud in the low register and produces a softer tone in the middle register. Given such differences, the composer may need to tailor the arrangement in light of the specific instruments used. Now suppose we have difficulty recruiting a flute and oboe player. Can we substitute a clarinet for the oboe? How about using two trumpets? Now consider a work written for two flutes. Since the timbres are matched, the composer does not need to pay as much attention to tailoring the arrangement. Moreover, we could easily substitute two violins or two oboes for the two flutes. In general, creating works for matched timbres makes it easier to substitute different instruments. Especially in earlier historical eras, the practical problems of recruiting specific instruments might have discouraged composers from using different timbres as a way of enhancing the perceptual independence of the parts.

A related consideration arises from the goal of balance. As long as the parts are performed on similar instruments or are restricted to human voices, no single part is likely to be much louder or more noticeable than another. Once heterogeneous timbres are adopted, all sorts of differences emerge. Pairing a flute with a trombone will introduce a significant discrepancy in acoustical power. Moreover, apart from loudness, some timbres are simply better able to grab the listener's attention; a human voice, for example, will nearly always command more attention than a musical instrument.[29] Similarly, a French horn is no match for an oboe; provided both instruments are equally busy, the oboe will typically dominate the listener's attention. Notice that the use of distinctive timbres seems to occur when the composer wants to draw attention to a particular line of sound against a background of other sounds. That is, timbral differentiation is often used to establish relationships between foreground and background.

Another plausible (and related) goal is that of achieving a homogeneous blend of sound. Working at Northwestern University, Gregory Sandell studied a number of perceptual factors influencing orchestral blend.[30] Sandell has noted that composers tend to use instrumental combinations that show a high degree of blend. It may be that homogeneous instrumentation merely provides the easiest way to achieve a blended sound. Although heterogeneous instrumentation would enhance the perceptual segregation of the voices, it seems that many composers have typically found the overall result less preferable than a uniform ensemble.

Source Location

One of the best generalizations that can be made about independent sound sources is that they normally occupy distinct positions in space. This suggests that location should provide strong streaming cues. Research by Pierre Divenyi and Susan Oliver at the Department of Veterans Affairs in Martinez, California, has shown that spatial location has a strong effect on streaming.[31] Even in reverberant environments, listeners are adept at identifying the physical location from which a sound originates.[32]

Drawing on this research, we can formulate the following principle:

10. Source Location Principle
Different source locations contribute to the perceptual independence of concurrent voices or parts.

Once again, musical practice appears to be equivocal. Some music does indeed make use of separate locations to segregate the musical sources. In the antiphonal works of Giovanni Gabrieli, groups of instruments or singers may be separated by 50 meters or more. In most music, when two or more instruments combine to play the same part, they are typically seated in the same acoustic neighborhood, as in the second violin section of an orchestra.

Other spatial effects suggest a different goal. In the case of barbershop quartet singing, common performance practice places the singers in very close physical proximity with the singers' heads positioned near a single location in space.[33] This arrangement implies that the musicians intend the voices to meld into a single cohesive auditory image rather than listeners forming independent images of the various singers' voices. This interpretation is supported by the frequent use of parallel motion in barbershop arrangements. Pitch glissandi are also common, a practice that is especially likely to cause the partials to be assembled into a single auditory image. In addition, all of the voices occupy a narrow pitch range that makes it more difficult for listeners to decipher the individual voices. The close pitch proximity explains why mixed choruses (female and male) are avoided in close harmony. Same-sex ensembles are the norm: all-male groups for barbershop style and all-female groups for the parallel style known as Sweet Adeline. All of these features suggest that the close-harmony genre aims at synthetic perceptions where the entire chorus evokes a single virtual source.

A number of musical works in the late twentieth century conspicuously manipulated spatial effects. In the case of instrumental works, spatial effects are often associated with unorthodox seating arrangements. In extreme cases, such as Iannis Xenakis's *Terretektorh*, eighty-eight instrumentalists are dispersed throughout the audience. In general, musicians have tended to avoid such wide separations because listeners often find themselves unable to hear much beyond the musician sitting next to them. In *Terretektorh*, however, Xenakis likely considered this perceptual goal less important than a sociological goal of attempting to break down the traditional dichotomy between performers and audience.

Xenakis's *Terretektorh* notwithstanding, performers rarely take advantage of the available opportunities to separate themselves in space. On most stages, the members of a string quartet cluster at center stage rather than positioning themselves at the maximum distances afforded by the stage.

There are many reasons why musicians might be reluctant to disperse themselves in space. During rehearsals, conversation is easier when the musicians are close together, and this arrangement might be preferred by habit. Another reason is that performers may have trouble hearing the other musicians over the sound of their own instruments. This problem of balance also arises for members of the audience. As sound sources become more widely separated, the musical effect for different audience members becomes more variable: the balance between instruments depends primarily on where the listener sits. Conversely, when all the sound sources are in close physical proximity, each sound source is placed on a more equal footing. As in the case of timbral differentiation, the traditional reluctance to employ spatial differentiation may arise due to problems in maintaining ensemble balance. Finally, musicians might simply feel more comfortable when they maintain a normal social distance among themselves.

In the case of electroacoustic music, spatial manipulation may prove more effective. Much electroacoustic music is experienced via stereophonic recordings in personal listening settings, minimizing the effect of listener location. In addition, spatial cues may be dynamic (trajectories) rather than static (fixed location). By using continually moving virtual sound sources, it may be possible to maintain a relatively homogeneous texture without unintended foreground and background effects.

Reprise

In this chapter, we have discussed four additional perceptual principles that can be used to enhance the segregation of auditory streams: *onset synchrony, limited density, timbral differentiation*, and *source location*. Compared with the six principles discussed in the previous chapters, these four principles appear to have had less impact on past music-making. In the case of onset synchrony, composers have commonly created music that is entirely contrary to the principle. I have suggested that the reason why these four auxiliary principles are often ignored is that they easily conflict with other goals that composers often pursue. For example, goals such as ensemble balance, lyric intelligibility, harmonic clarity, or rhythmic energy may require some compromises with the goal of auditory stream segregation.

Whenever more than one goal is being pursued, there are often saddle-points of stability where an optimum compromise is reached. I propose that this may be a useful way to view common musical textures. That is, different textures, like close harmony, tune and accompaniment, pseudo-polyphony, polyphony, monophony, heterophony, and homophony, may represent different combinations of goals. Since different scene-analysis principles have different consequences for the composer pursuing other goals, some of the principles have the appearance of compositional options. However, whether a given work conforms or does not conform to a given principle I think is primarily a consequence of the composer's pursuing other goals at different times. In later chapters, we will see more compelling examples of compromises arising from competing compositional goals— such as in the development of parallel harmonies in modern African choral music.

9   Embellishing Tones

Four-part harmony sometimes consists of just a succession of block chords, such as found in many hymns. But even with the most lumbering harmonic passages, it is common for composers to spice things up by adding embellishing tones.

Different theorists have used different labels for these tones, including *nonchordal tones*, *nonharmonic tones*, *figuration tones*, *unessential tones*, and *embellishing tones*.[1] The term *nonchordal tone* recognizes that most of these tones do not belong to a given chord. The term *nonharmonic tone* draws attention to the fact that these tones do not change the harmonic function or progression of a passage. The term *figuration tone* implies an ornamental or decorative function. The term *unessential tone* suggests that these tones have a secondary importance and may be discarded without changing the basic character of a musical passage. Of the various terms in use, I prefer *embellishing tone* since some of the tones may be members of a chord or participate in defining a harmonic function. At the same time, the term *embellishment* seems less derogatory than *unessential*.

Music theorists have produced a widely accepted classification scheme for embellishing tones. The tones are categorized according to five criteria:

1. Whether the tone belongs to a sounding chord
2. Whether the tone appears in a metrically accented position
3. Whether the tone is approached by a scale step
4. Whether the tone is resolved or left by step
5. The pitch contour in which the tone is placed (e.g., up-up, up-down, down-up)

Embellishing tones include passing tones, neighbor tones, appoggiaturas, escape tones, anticipations, pedal tones, changing tones, retardations, suspensions, repetitions, and arpeggiations. Figure 9.1 illustrates the main embellishing tones found in Western music.

Theorists have suggested a number of possible functions or rationales for embellishing tones. A common theme in most accounts is that embellishing tones add interest to a musical texture. This interest might arise, for example, by adding dissonance, creating or heightening a feeling of expectation, or drawing attention to neighboring structural tones.

Working at the University of Pennsylvania, the renowned theorist Leonard Meyer suggested that embellishing tones can be analyzed according to their interplay with the listener's expectations.[2] The flow of musical events can be characterized in terms of the generation of musical implications and the subsequent fulfillment or thwarting of expected outcomes.[3] Terms such as *suspensions*, *retardations*, and *anticipations* provide telling descriptive labels for the corresponding psychological experiences. Inspired by Meyer's perspective, in my book, *Sweet Anticipation*, I provided detailed psychological analyses of various embellishments.[4] However, there are several other interesting approaches to understanding the role of embellishing tones in music.[5]

In 1986, James Wright at McGill University suggested that the presence of embellishing tones might also contribute to the perceptual independence of the individual parts.[6] Twenty years later, I carried out a study that tested a series of hypotheses arising from his idea.[7] My study looked in detail at a sample of 50 chorale harmonizations by J. S. Bach. For each chorale harmonization, I created a "twin" where the embellishing tones were removed in a way that maintained the underlying chord progression. I then compared the original (embellished) harmonization with its unembellished twin. The presence of the embellishing tones has at least four effects:

1. The addition of embellishing tones decreases the average melodic interval size within each voice. That is, embellishments (on average) result in closer pitch proximity. Of course, this is not surprising to musicians: the most common embellishment is the passing tone, and passing tones often connect two chords where the voice would normally jump the interval of a third. Embellishments that increase the pitch distance (such as appoggiaturas and escape tones) occur much less frequently.

2. The addition of embellishing tones increases the overall degree of asynchrony. Once again, this is hardly surprising. Apart from pedal tones, most embellishments add a tone with its own onset moment.

Embellishing Tones

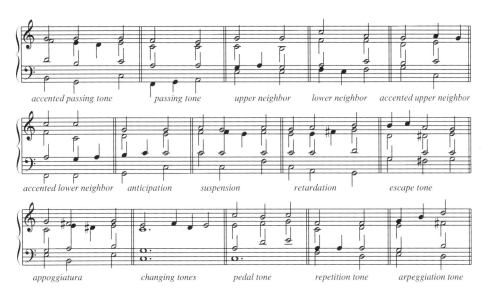

Figure 9.1
Illustrations of various embellishing tones. Examples are drawn from Bach chorale harmonizations.

3. The addition of embellishing tones decreases the amount of similar and parallel motion between concurrent parts. This is musically less obvious, but the pitch contours that result from the presence of an embellishment typically increase the amount of oblique and contrary motion. Said another way, embellishing tones reduce semblant motion or pitch co-modulation.
4. The fourth finding relates to which voice receives an embellishment. The most likely voice to be embellished in a sonority is a voice that currently duplicates a pitch class. This is illustrated in figure 9.2. In both examples, the initial chord is the same: E4, C4, G3, and C3. Notice that the C is doubled between the alto and bass. Any octave interval has the potential to evoke harmonic fusion, so if a composer were attempting to reduce the possibility of inadvertent fusion, then it would make sense to introduce an embellishing tone in one of the doubled voices. In figure 9.2a, a passing tone is introduced in the soprano (a voice whose pitch class is not doubled). In figure 9.2b, a passing tone appears in the bass (a voice whose pitch class *is* doubled). It turns out that Bach is much more likely to introduce an embellishing tone in a voice whose pitch class is doubled, as in figure 9.2b.[8] That is, Bach's use of embellishing tones is consistent with efforts to minimize or prevent harmonic fusion.

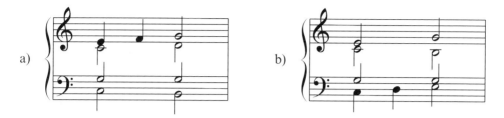

Figure 9.2
Illustration of two different ways of deploying embellishing tones. In both examples (a) and (b), the pitch C is doubled between the bass and alto voices. In example (a), the passing tone is assigned to a voice that does not double a pitch class. In example (b), the passing tone is assigned to a voice that does double a pitch class. The second example is more characteristic of J. S. Bach's musical practice and is consistent with efforts to minimize tonal fusion.

Of course there are other possible reasons that Bach might do this. For example, he might simply have wanted to maintain the presence of all of the chord factors (root, third, and fifth) throughout the beat. Accordingly, it makes sense to embellish one of the voices that duplicates an existing pitch class.

My four observations don't prove Wright's idea that embellishing tones are employed primarily in order to enhance the perceptual independence of the parts. All we can say is that Bach's compositional practice is consistent with this goal. Specifically, the presence of the embellishments is consistent with four core principles in forming independent auditory streams: the pitch proximity principle, the pitch co-modulation principle, the onset synchrony principle, and the harmonic fusion principle.

Taken at face value, these observations suggest some supplementary voice-leading preference rules. Let's start by linking together the pitch proximity, pitch co-modulation, and onset synchrony principles:

E9&E13&E15. *Stream segregation is enhanced by asynchronous sounds that create nonsemblant proximate pitch contours.*

From this we can infer a recommendation that encourages the use of embellishing tones:

[PR26]. **Embellishments Rule**. *It is preferable to interpose embellishments (like passing tones) between successive sonorities.*

Not all types of embellishing tones are equally good at preserving close pitch proximity. So some types of embellishments should be preferred over others:

[PR27]. **Embellishment Preference Rule**. *Prefer embellishments using step motion (like passing tones, neighbor tones, suspensions, and retardations) or no motion (pedal tones, repetitions, and anticipations). Use embellishments involving leaps (like escape tones, appoggiaturas, and arpeggiations) less often.*

Recall the earlier *leap away* rule (preference rule 16). This would suggest that escape tones, appoggiaturas, and arpeggiations should preferably appear in outer voices and should jump away from the texture. In Bach's chorale harmonizations, the majority of leaping embellishments do indeed occur in the soprano or bass voices, and 90 percent of those jump away from the other voices rather than toward them. Once again, this practice is consistent with efforts to avoid listeners becoming confused about the streaming of the individual parts.

Finally, we can add a fourth principle to the three linked above. Including the harmonic fusion principle:

E5&E9&E13&E15. *Stream segregation is enhanced by asynchronous sounds that create nonsemblant proximate pitch contours where harmonically fused intervals are minimized.*

From this combination of four principles, we can infer a recommendation regarding pitch-class duplication:

[PR28]. **Doubled Pitch-Class Rule**. *Prefer adding embellishments to voices that double the pitch class of another concurrent voice.*

Of course, not all part-writing contains embellishing tones. Many hymns, for example, contain no embellishments at all. In the discussion of onset synchronization in chapter 8, I suggested that other considerations might take a higher priority—for example, preserving the intelligibility of sung lyrics, establishing an energetic rhythmic character, and promoting the synthetic perception of chords. Such considerations might discourage composers from using embellishing tones.

Turn Taking

When I was a teenager studying music, I participated in a weekend retreat with some fellow students. The purpose of the excursion was to heighten our awareness of sound, so all of us spent the entire weekend blindfolded. The surroundings were unfamiliar, so simple activities like preparing meals proved to be challenging. What I most recall about the experience, however, was what happened when people were sitting around in a room having a conversation. There might be ten people in the room, but perhaps only four people would be actively talking. It was difficult to keep track of who was present. Those who didn't say anything for five or ten minutes effectively faded from the room. After half an hour of saying nothing, no one would know that you were there. It was a case of *out of sound, out of mind*.

When you add an embellishing tone to some voice, it draws attention to that voice. A passing tone in the alto effectively reminds listeners of the alto's existence. Could it be that embellishing tones are distributed among the voices as a sequential way of reminding listeners of the existence of each voice? In my study of embellishing tones in Bach's chorale harmonizations I also analyzed how the embellishments were assigned to the various voices. Figure 9.3 shows the proportion of embellishments found in the four voices in Bach's chorale harmonizations.

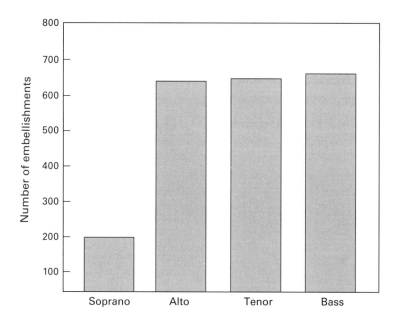

Figure 9.3
Frequency of occurrence of embellishing tones in chorale harmonizations by J. S. Bach. Embellishing tones occur with nearly identical frequency in the bass, tenor, and alto voices. Relatively few embellishments are found in the soprano. Source: Huron (2007).

As can be seen, the alto, tenor, and bass voices have roughly equivalent numbers of embellishments. The soprano, however, shows considerably fewer embellishments. Of course, as the highest voice, the soprano already enjoys an increased noticeability due to the high-voice superiority effect. But the relative paucity of embellishments in the soprano probably has a more prosaic origin. The chorale harmonizations are based on hymn tunes popular in Bach's day; the melodies are given. So the relative absence of embellishments in the soprano may be because Bach resisted tinkering with well-known melodies.

In tracing the assignment of embellishments to different voices, I was surprised to discover evidence of a sort of musical turn taking. Not only did Bach distribute the embellishments fairly evenly among the voices, but he also assigned embellishments with a sensitivity to which voice had been ignored the longest. Suppose that Bach introduced an embellishment in the tenor voice, then later an embellishment might appear in the bass, then the tenor again, followed by the soprano. With better-than-chance results, one could predict that the next embellishing tone would appear in the alto. That is, I found a strong statistical tendency to assign the next embellishment to the voice that had gone the longest without an embellishment. As in my blindfolded retreat experience, the most important reminders are of those who have been out of mind for the greatest length of time. If we want to keep track of who is present, we most need to hear from the voices that have been quiet the longest.

Once again, the fact that Bach's deployment of embellishing tones is consistent with a scheme for restoring attention to individual streams does not prove that this was Bach's intention or goal. All we can say is that this practice is consistent with the many other stream-enhancing behaviors evident in Bach's music. In light of this analysis, we might propose one more preference rule related to embellishing tones. First is the pertinent psychological principle:

11. Attention Principle
When competing with other concurrent auditory streams, streams tend to fade from attention. A listener's attention is drawn to isolated sound events, so the existence of a given auditory stream can be enhanced by introducing a solo onset event.

From this we can infer:

[PR29]. **Embellishment Refresh Rule**. *Prefer to add embellishments to voices that have not been embellished recently.*

Reprise

Why do embellishing tones exist? Embellishments might serve a number of functions, including adding dissonance, creating or heightening expectations, drawing attention to neighboring structural tones, or simply adding interest to a musical texture. In this chapter we have pointed out that common embellishments are consistent with a number of principles of auditory scene analysis, and that the addition of embellishing tones improves voice independence in at least five ways. First, most embellishing tones improve pitch proximity by reducing the number of melodic leaps. When leaps are introduced, they tend to occur in outer voices and leap way from the texture, consistent with minimizing stream competition for ensuing pitches. Second, the addition of embellishing tones tends to reduce onset synchrony. Third, the presence of embellishing tones decreases the amount of similar and parallel motion between the parts. Fourth, embellishments tend to be assigned to voices that duplicate pitch-classes, and so reduce the likelihood of harmonic fusion. Finally, embellishing tones tend to be assigned to different voices in a way that draws attention to seemingly forgotten or ignored parts. Embellishments may serve other functions as well, but in practice their presence contributes to the perceptual independence of the concurrent parts, consistent with the goal of voice leading.

10 The Feeling of Leading

In his classic book on auditory scene analysis, Albert Bregman distinguished two approaches that brains use to assemble auditory scenes: *bottom-up* and *top-down*.[1] The bottom-up approach groups partials together according to shared features (like onset synchrony and harmonicity). In the top-down approach, the brain starts with one or more intuitions about how the acoustic scene is organized and then looks for evidence consistent with these preexisting intuitions. So far, our discussions have focused exclusively on the bottom-up approach.

A useful analogy for these different approaches can be found in the assembling of a jigsaw puzzle. A bottom-up approach emphasizes finding individual puzzle pieces that seem to fit together. A top-down approach might sort the puzzle pieces according to the overall picture on the puzzle box. You might sort all of the sky pieces into one area, the foliage pieces into another area, and so on. As with a jigsaw puzzle, auditory scene analysis is most effective when both bottom-up and top-down approaches are employed together.

In auditory scene analysis, the bottom-up approach lets the scene emerge from the affinity of the sounds themselves: pitch proximity, pitch co-modulation, location, and so on all contribute evidence that the auditory system uses to assemble plausible auditory images and streams. In the top-down approach, by contrast, the brain attempts to jump-start the process by arranging the incoming information to fit some prior assumptions about how the acoustic world should be. For example, when we hear people singing a familiar song like "Happy Birthday," we expect a particular sequence of pitches. These expectations can increase the likelihood of hearing the expected pattern, especially in a complicated acoustic scene such as when there is a lot of background noise. In parsing the auditory scene, the top-down approach emphasizes more contextual or global features than the bottom-up approach.

One aspect of the top-down approach is *intersensory agreement.* In assembling a picture of the world, we don't rely solely on our ears. Brains coordinate information from all of the senses with the aim of forging some consensus. In parsing auditory scenes, our brains prefer that our ears agree with our eyes. An example of this intersensory process can be found in the art of ventriloquism. A ventriloquist is able to speak with little or no mouth movement. At the same time, the ventriloquist coordinates the mouth movements of a dummy. The brain has to reconcile the visual information with the auditory information. At a live performance, auditory cues may suggest that the sound originates in the vicinity of the ventriloquist's mouth, but the visual cues suggest that the action is taking place in the region of the dummy's mouth. In this case (and most cases of spatial conflict) the visual system wins out over the auditory system: we "hear" the sounds coming from the dummy rather than from the human performer. When it comes to judging location, the brain assumes that the visual information is more reliable than the auditory information.

Such intersensory coordination is not limited to the unusual circumstance of the ventriloquist. It happens all the time. When we see three clarinet players walk onto a stage, this visual information might prepare us to hear three musical lines, each with the timbre of a clarinet. Having observed the clarinet players, we would be surprised to hear the sound of a single accordion instead.

While intersensory coordination is useful, the principal tool for top-down scene analysis is *expectation.*[2] Research has established that future expectations are integral to visual and auditory perception.[3] Predicting the future is biologically useful: our ability to anticipate future events makes it easier for us to take advantage of opportunities and to sidestep dangers. By anticipating future sounds, we provide the brain with a head start in recognizing and interpreting the incoming sounds. This head start allows the brain to process expected sounds faster than unexpected sounds. The same principle applies to vision. As your eyes follow along this text, your brain is anticipating what words will occur next—with the consequence that your brain will figure out the meaning of the sentence more rapidly.

The facilitating effect of expectation on the speed of processing has been observed in many different animals and with all kinds of visual and auditory stimuli. In the case of melodies, the facilitating effect of expectation has been demonstrated in experiments by Bret Aarden at the Ohio State University.[4] In Aarden's experiments, listeners were asked to track the pitch contour of a melody and press one of three buttons after the onset of every tone in the melody. The buttons indicate whether the pitch (1) has gone up, (2) has gone down, or (3) has remained the same. When hearing a leading tone, for example, listeners are quite fast to press the "up" button if the ensuing pitch rises to the tonic. If the leading tone is followed by a lower pitch, however, listeners are relatively slow to press the "down" button. The difference in reaction times is symptomatic of the fact that listeners expect the leading tone to rise, which is indeed what happens most often in tonal music.

The role of expectation in hearing musical streams is nicely illustrated by so-called interleaved melodies. Working at the University of California, Los Angeles in the 1970s, psychologist Jay Dowling carried out a series of studies in which he interleaved two well-known melodies. He would begin by playing the first pitch of melody A, followed by the first pitch of melody B, followed by the second pitch of melody A, followed by the second pitch of melody B, and so on until the end of both melodies. The question Dowling asked his listeners was, "Can you identify either of the two melodies present?" Figure 10.1 shows an example of two interleaved melodies.[5]

On first hearing, listeners are typically clueless and are unable to decipher either of the two component melodies. But Dowling would repeat the stimulus; with each repetition, he would transpose the melodies so they were moved farther apart. That is, the pitches for one melody he would transpose upward, while the pitches for the other melody were transposed downward. Figure 10.2 shows the same interleaved melodies in figure 10.1, but one melody has been transposed up five semitones while the other melody has been transposed down five semitones. As the melodies are

Figure 10.1
Two interleaved melodies: the odd-numbered pitches form the melody "Frère Jacques." The even-numbered pitches form the melody "Twinkle, Twinkle, Little Star." *Source*: After Dowling (1967).

Figure 10.2
The same interleaved melodies as in figure 10.1 transposed. "Frère Jacques" (odd-numbered pitches) has been transposed upward, while "Twinkle, Twinkle, Little Star" (even-numbered pitches) has been transposed downward, increasing the likelihood of hearing out the two melodies.

transposed farther and farther apart, listeners at some point would suddenly hear the melodies jump out; that is, they could hear (and identify) the two melodies used to create the interleaved tone sequence.

Dowling's interleaved melodies demonstrate why the crossing of parts can be so bad: listeners tend to hear pitches as connecting to the nearest subsequent pitch. When both melodies occupy the same pitch region, it can be very difficult to hear any melody other than the sequence of pitches formed by the alternating pitches from both melodies.

In carrying out his studies, Dowling also observed that once a person knew what to listen for, it became much easier for him or her to hear out the constituent melodies making up the interleaved tone sequence. For listeners who had already done the task once, less transposition was required before they could hear the tone sequence as two independent melodies. In short, top-down knowledge (knowing what tune to expect) facilitated the separating out of the auditory streams. Auditory scene analysis can be influenced by higher-level mental processes.

Once a listener knows that "Frère Jacques" is one of the interleaved tunes, he or she can actively attempt to detect its presence. That is, by an act of will, a listener can sometimes change whether a tone sequence is heard as one stream or two streams. Several researchers have observed this same phenomenon in different acoustic circumstances. For example, this willful "hearing-out" can sometimes occur when listening to the individual harmonics of a complex tone. When a harmonic tone is sustained for many seconds, it is sometimes possible to hear-out a specific partial. What was previously perceived as a single complex tone (a single auditory image) becomes transformed into two images (the isolated partial, plus the rest of the complex tone). In auditory scene analysis, this sort of hearing out is referred to as *analytic listening*.[6] Analytic listening is perhaps the ultimate example of a top-down approach to auditory scene analysis. Here the listener consciously chooses how to parse the scene: "I want to hear this as a single stream." "I want to hear this as two streams."

It bears emphasizing that analytic listening is very limited. Our conscious intentions cannot dictate how the auditory system behaves. Simply saying, "I want to hear this violin melody as a sequence of parallel major chords consisting of the separate fourth, fifth, and sixth harmonics," will not necessarily make it happen. In fact, it is highly unlikely. With practice, musicians can often develop the imaginative skills that make it possible to perform certain (restricted) acts of analytic listening. But even the most practiced musician cannot force the auditory system to parse the auditory scene any arbitrary way.

While analytic listening involves an act of will, most top-down processing is actually the consequence of *unconscious* expectations. If pitch successions conform to expectations then the corresponding stream organization is apt to be facilitated. As Dowling demonstrated, how we parse an auditory scene depends to some extent on what we think will happen next.[7] Listening to music is not simply a passive response to sequences of sounds. Listeners anticipate what will happen next. We project ahead: we listen into the future.

Time's Arrow

Suppose we took a passage of well-written four-part harmony and played it backward. Would the modified music still obey the part-writing rules? Figure 10.3 provides an example. It shows two phrases from a four-part hymn in both original and backward (retrograde) arrangements.

With a moment's thought, we can identify the rules that would continue to be obeyed in the backward rendition. Of course, the ranges of the voices would remain identical in both the forward and backward versions. The spacing of chordal tones would be the same, as would the avoidance of unisons and doubled leading tones. Also, if the part-writing avoids crossed parts, then both forward and backward versions would be fine. Part overlapping would also be identical in both versions. The preservation of common tones would also be insensitive to direction. Similarly, if the part-writing doesn't have any consecutive fifths or octaves, then whether the music is played forward or backward doesn't matter; there wouldn't

Figure 10.3
Final two phrases of the four-part hymn "St. Flavian": (a) normal; (b) same passage played backward. Are any part-writing rules transgressed in the backward version?

be any consecutives in either direction. The traditional injunction against augmented melodic intervals is also insensitive to time's arrow: a melodic interval is the same distance whether we go forward or backward.

There are a couple of part-writing rules, however, that aren't symmetrical with respect to time. For example, if a particular part moves to the nearest chordal tone in the forward direction, it doesn't necessarily follow that it will move to the nearest chordal tone when played backward. Nevertheless, the average melodic pitch movement would be the same. That is, the overall pitch proximity would be identical.

Only one part-writing rule is clearly sensitive to direction: the direct (or hidden) octaves rule. The traditional rule tells us that we need to be careful about how we approach an octave—not how we continue after an octave interval. In a sample of ten four-part hymns, I found that reversing the order of the chord progressions added just two part-writing transgressions over a total sequence of 332 chords. Both of the transgressions were direct octaves. Apparently there isn't a great deal of difference between forward and backward part-writing: if the part-writing is good in one direction, playing it backward will do little harm.

Obviously playing a musical work backward produces a very different listening experience. To get a sense of how different, I recommend playing the original and retrograde passages shown in figure 10.3. For most passages, the backward version will sound much less musical than the forward version. Yet the backward version doesn't exhibit dramatically worse part-writing, at least when the music is assessed from the perspective of the traditional rules. Much of the blame lies with the mangled harmonic progressions, including failures to properly resolve harmonic dissonances and the metric misalignment of stable harmonies. But apart from the harmony, the moment-to-moment connections in the various parts still sound bad. Evidently, there is more to good part-writing.

Going Somewhere: Voice Leading

Music often conveys a feeling of forward momentum, a feeling of going somewhere. Played backward, most musical passages have an aimless, meandering quality that makes the sounds seem pointless. As we have already seen, listeners don't simply hear a succession of sounded events; they anticipate what will happen next.

Some chords and tones evoke a strong sense of tending, yearning, or leading. In most musical contexts, the aptly named leading tone commonly evokes a strong feeling of wanting to rise upward to the tonic. Similarly, in the majority of musical contexts, the lowered leading tone (the so-called subtonic) evokes a downward urge leading to the submediant. By contrast, the tonic pitch commonly tends to evoke a sense of stability and repose. I propose that this dynamic sense of tones wanting to go somewhere is what distinguishes voice leading from part-writing. Music played backward can exhibit good part-writing but poor voice leading.

So what accounts for these feelings of "tending"? Over the past decade, significant advances have been made in our understanding of auditory expectation. Important research has been carried out by Glenn Schellenberg, Mark Schmuckler, Barbara Tillmann, William Thompson, Bret Aarden, Elizabeth Margulis, Lola Cuddy, Eugene Narmour, and others.[8] In many circumstances, listeners do indeed form strong expectations about what will happen next. Moreover, the research shows that listeners develop several forms of expectation. When we are familiar with a given musical work, we form specific expectations of how the work goes. These sorts of piece-specific expectations are referred to as *veridical expectations*. Independent of these expectations, listeners also develop more general expectations of "how music in general goes" or "how music in this style goes." These broader expectations are referred to as *schematic expectations*.[9] Amazingly, the research suggests that veridical and schematic expectations have different neurological origins.[10]

When we listen to music, schematic expectations are always present, even when we are not familiar with the music. When the music is familiar, veridical expectations are added to the experience; that is, for familiar passages, schematic and veridical expectations operate simultaneously. It is possible for a musical passage to be schematically surprising but veridically expected. Similarly, it is possible for a passage to be veridically surprising but schematically expected. Listeners familiar with Western music will have heard the *V–I* chord progression hundreds of thousands of times. In most contexts, when we hear a *V* chord, our schematic expectation is that it will be followed by a *I* chord. This means that a *V–vi* (deceptive) progression will be schematically surprising even when we are intimately familiar with the musical work. When we are familiar with the work, our veridical expectation anticipates *V–vi*, but our general schematic expectation that *V* is usually followed by *I* will still be violated, and it is this schematic violation that accounts for the "deceptive" quality that we continue to feel.[11]

Where do our expectations come from? A wealth of research has established that expectations are formed from simple exposure. The more times we hear some particular sequence, the stronger the expectation is that that sequence will unfold as it has in the past.[12] A good illustration of the role of exposure can be found when listening to albums or fixed playlists on a portable personal music player. With repeated listening, a given playlist soon becomes familiar. Hearing the end of one piece, you will have a quite specific expectation of what the next musical selection is and how it starts. Even if you assemble an entirely arbitrary playlist, it doesn't take many repetitions before your brain comes to anticipate what the next piece is. This experience highlights three facts about expectation. First, auditory expectations are learned from simple exposure. Second, we can learn to expect anything following anything else (the learned sequence of events can be completely arbitrary). And third, expectations are strongest over a short time frame: we often aren't sure what piece will happen next until the current piece is about to end.

This pattern of learning by exposure is referred to as *implicit learning* (or *statistical learning*), and extensive experimental evidence suggests that implicit learning provides the foundation for musical expectations.[13] This research is reviewed at length in my book, *Sweet Anticipation: Music and the Psychology of Expectation*, so I won't summarize it here.

Table 10.1
Probabilities of successive diatonic pitches (major mode only)

Antecedent	→$\hat{1}$	→$\hat{2}$	→$\hat{3}$	→$\hat{4}$	→$\hat{5}$	→$\hat{6}$	→$\hat{7}$	→rest
$\hat{1}$→	3.4	2.8	2.0	0.2	1.3	0.8	2.3	3.7
$\hat{2}$→	4.2	2.6	3.3	0.7	0.8	0.2	0.6	1.5
$\hat{3}$→	1.6	4.9	3.1	2.6	2.4	0.3	0.0	2.4
$\hat{4}$→	0.1	1.3	4.1	1.5	1.7	0.4	0.1	0.5
$\hat{5}$→	2.6	0.5	2.9	3.7	4.8	2.1	0.4	2.3
$\hat{6}$→	0.2	0.2	0.1	0.3	3.6	1.3	0.9	0.4
$\hat{7}$→	2.0	0.5	0.0	0.0	0.3	1.3	0.4	0.3
rest→	2.0	1.1	1.6	0.7	3.1	0.5	0.2	—

If listeners form expectations on the basis of past exposure and if we want to know what people are likely to expect next, a straightforward approach is to look at the kinds of sequential patterns that are commonplace in music. Consider the case of successive scale degrees. Table 10.1 shows the frequency of occurrence for successive scale tones in a sample of Western vocal melodies in the major mode. The table identifies the percent likelihood that some tone (the antecedent tone) is followed by another tone (the consequent tone). Antecedent tones are shown in rows and consequent tones in columns. For example, the table shows that the tonic pitch is followed by a repetition of the tonic pitch in 3.4 percent of cases. That is, of all of the possible tone pairs, 3.4 percent involve two successive tonic pitches.[14]

The most probable melodic event in this musical sample is $\hat{3}$ followed by $\hat{2}$. With a calculated probability of 4.9 percent, almost one in twenty pitch transitions involves moving from $\hat{3}$ to $\hat{2}$. Repetition of the dominant pitch occurs nearly as frequently, while the third most common melodic succession is from $\hat{2}$ to $\hat{1}$.

Figure 10.4 displays some of the data from table 10.1 in graphic form showing the probabilities of various scale-degree successions. Once again, remember that these probabilities are for melodies in major keys only. In this illustration, the thickness of the lines is proportional to the likelihood of occurrence. Not all melodic transitions have been graphed; connecting lines have been drawn only for those transitions that have a probability

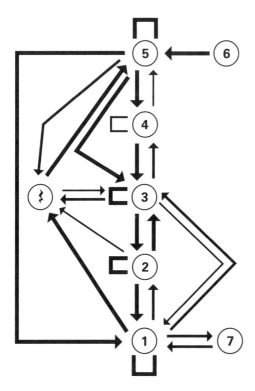

Figure 10.4
Schematic illustration of scale-degree successions for major-mode melodies from Huron (2006b). The thickness of each connecting line is proportional to the probability of melodic succession. The quarter-rest symbol signifies any rest or the end of a phrase. Lines have been drawn only for transitions that occur more than 1.5 percent of the time.

greater than 1.5 percent. A number of patterns can be seen in figure 10.4. Notice the prevalence of step motion and repeated pitches in accordance with the pitch proximity principle. Also notice that descending steps are more common than ascending steps. The backbone of the diagram is a descending sequence from $\hat{5}$ to $\hat{4}$ to $\hat{3}$ to $\hat{2}$ to $\hat{1}$.

Notice that several pitch pairs in figure 10.4 are asymmetrical. For example, the sixth scale degree is much more likely to lead to the fifth scale degree than vice versa. Similarly, the fifth scale degree is more likely to move to the fourth scale degree than the other way around. Some asymmetries are counterintuitive. Notice that the arrows between $\hat{1}$ and $\hat{7}$ indicate

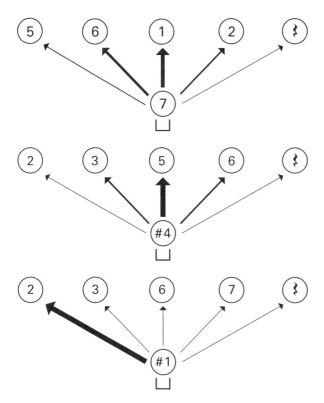

Figure 10.5
Schematic illustration of scale-degree successions for $\hat{7}$, #$\hat{4}$, and #$\hat{1}$ for major-key melodies. The thickness of each connecting line is proportional to the probability of melodic succession. The quarter-rest symbol signifies any rest or the end of a phrase.

that $\hat{1}$ to $\hat{7}$ is more probable than $\hat{7}$ to $\hat{1}$. This might seem odd, but bear in mind that this data represent absolute likelihoods rather than relative likelihoods. The tonic pitch simply occurs more often than the leading tone.

More important than the absolute probabilities are the relative probabilities, and these are more easily grasped if we isolate individual scale degrees. Figure 10.5 focuses on three tones: the leading tone ($\hat{7}$), the raised subdominant (#$\hat{4}$), and the raised tonic (#$\hat{1}$). In each case, the five most probable continuations are plotted. Once again, the thickness of the lines is proportional to the probability of occurrence. In the case of the leading tone, the general affinity for $\hat{1}$ is obvious. In absolute terms, there are more instances of $\hat{1}$ going to $\hat{7}$ than $\hat{7}$ going to $\hat{1}$. But $\hat{7}$ is more likely to lead to

$\hat{1}$ than $\hat{1}$ is to $\hat{7}$. In most contexts, the fate of $\hat{7}$ is more strongly linked to $\hat{1}$ than vice versa.

What $\hat{7}$, $\sharp\hat{4}$, and $\sharp\hat{1}$ share in common is that a single continuation pitch generally dominates all other possibilities. It is not inappropriate to call these tones "tendency tones." A few years ago I carried out a survey where I asked musicians to describe how the different scale degrees "feel."[15] The musicians described $\hat{7}$, $\sharp\hat{4}$, and $\sharp\hat{1}$ by using such terms as *unstable, pointing, restless*, and *itchy*. In other words, the subjective experience of *tendency* is strongly associated with the probability of one tone following after another.

It is important to understand that tendencies are strongly context dependent. For example, in the context of a cadential 6–4 chord, the tonic pitch will be more strongly attracted to the leading tone than vice versa. In this case, it is the tonic pitch that is the tendency tone rather than the seventh scale tone. The important point is that these "leading" tendencies are learned from experience. Sensitivity to both the contexts and the tendencies arise from implicit learning through exposure to lots of music.

In effect, the auditory system is constantly asking, "What is most likely to happen next given the current situation?" By "current situation" we mean not just the isolated individual pitch, but also the melodic and harmonic context, and even the style of the music. The auditory system forms pretty good hunches about what to expect, and when experience suggests that there is only one good possibility, the auditory system passes these hunches on to consciousness in the form of feelings like yearning, tending, and leading.

When a musical passage is played backward, the pitch sequences are not likely to conform to schematic or veridical expectations. As a result, the sense of momentum, direction, flow, or inevitability will be lost. This gives the backward music its aimless quality. At the same time, any intentional moments of surprise created by the composer will no longer be properly prepared or resolved. Improbable tone sequences will simply seem arbitrary and weird.

Incidentally, these experiences are commonplace when listening to music from an unfamiliar culture. When a Western listener hears Indonesian gamelan music for the first time, the lack of familiarity with the appropriate *slendro* or *pélog* scale schemas can lead to the impression that successive tones simply wander around aimlessly. A similar experience is common for inexperienced listeners hearing Gregorian chant. A lack of familiarity of the tendency relationships in, say, Phrygian mode, can lend a meandering feeling to the chant that is absent for more seasoned listeners of medieval music.

The loss of a sense of musical direction in backward renditions of music can hamper the perception of independent parts. As we've seen, a retrograde rendition of a musical passage may violate few bottom-up principles of part-writing, but it can wreak havoc with the top-down expectations that assist in following a multivoice texture. Like the avoidance of parallel fifths, the sense of leading contributes to our ability to decipher the individual parts. For example, the expectation that, in a dominant–tonic cadence context, the leading tone will move to the tonic augments the bottom-up sense of connection arising from their close pitch proximity.

Imitative Part-Writing

So far we have distinguished just two forms of expectations: *schematic* (how music in some style goes) and *veridical* (how a familiar work goes). However, these are not the only kinds of expectations. Another kind is so-called *dynamic-adaptive* expectations.[16] These are expectations that arise on the fly. By way of illustration, read the following sequence of numbers aloud:

5–9–8–3–5
5–9–8–3–5
5–9–8–3–5

After the second repetition, you could probably repeat the sequence a third time without having to look at the next line. That is, the sequence was retained in short-term memory. Like your telephone number, if you repeat the sequence often enough, it will be transferred to long-term memory (where it may stay with you for the rest of your life). But if you don't rehearse the sequence, you are unlikely to retain it beyond the end of this chapter.

It is not just numbers or words that can be retained in short-term memory. Pitch sequences can also be retained in the same way. Suppose I ask you to sing the sequence notated in figure 10.6 three times.

After the second repetition, you could probably repeat the passage without having to look at the notation. As with the number sequence, after just one or two repetitions, the pattern was taken into short-term auditory

Figure 10.6

Figure 10.7

memory. Moreover, this pattern is now available for predicting the completion of a partial statement. With just a brief exposure, your auditory system knows how to complete the tone sequence in figure 10.7.

This musical pattern can now serve as a top-down template that facilitates following any auditory stream that includes the pattern. Even if this tone sequence is embedded in a complex texture containing other concurrent parts, prior exposure to this pattern (earlier in the piece) increases the likelihood that you will be able to follow the particular musical stream containing it.[17] In traditional Western polyphony, such recurring tone patterns are commonplace. In fact, the most common form of polyphonic writing, imitative polyphony, is based precisely on the use of such repetitive segments.

Perhaps the best-known imitative polyphonic form is the *fugue*. The most characteristic part of the fugue is the exposition in which a single theme (as *subject* or *answer*) is restated as each successive voice enters. Short-term memory retains the initial statement, and our ability to predict future statements helps facilitate following the individual lines as they appear.

The most extreme imitative polyphonic form is the *canon*, where each voice or part presents the same material, only offset in time (and sometimes offset in register). Canonic rounds like "Row, Row, Row Your Boat" have all the voices repeat the melody at the same pitch height. This means that there are no range differences to help distinguish the voices: part-crossings are commonplace and pitch-proximity violations are rampant. Like Dowling's interleaved melodies, the best hope listeners have for following the component lines is familiarity with the melody.

Fugues and canons aside, the tendency to repeat distinctive melodic patterns is evident throughout the vast repertoire of polyphonic works in Western musical culture. Imitative polyphony is the norm, even for works that are not fugues or canons.

Pause

Let's pause now and summarize the main points of this chapter so far. We've learned that top-down information can supplement bottom-up information in the processing of auditory scenes. Knowledge of how a musical line goes will assist the auditory system in tracking a stream. The principal tool for top-down processing is expectation. When we correctly anticipate what will happen next, auditory scene analysis gets a helping hand. Three forms of expectation were described in this chapter. *Schematic* expectations represent general knowledge about "how this kind of music goes." *Veridical* expectations are specific forms of knowledge about "how this particular (familiar) piece goes." Finally, there are *dynamic-adaptive* expectations that arise as listeners anticipate what might happen next based on what they've heard of the piece so far.

On the basis of research on auditory expectation, we can now propose the following principle:

12. Expectation Principle
The (top-down) parsing of auditory scenes is facilitated by accurate listener expectations. A listener will be better able to segregate concurrent musical lines when the composer follows conventional musical patterns (schematic expectations), quotes or alludes to familiar melodies (veridical expectations), or employs repetitive or imitative melodic patterns (dynamic-adaptive expectations).

Apart from the facilitating effect on stream segregation, accurate prediction also evokes a sense that the sequence is going somewhere. As listeners, we may experience one event as leading to another; some sounds yearn to be followed by another sound; the sequence generates feelings of momentum, direction, and even a sense of inevitability. (We also recognize when a composer has taken a detour, paused, jumped ahead, or stopped.) Conversely, when we are unable to predict how a series of sounds unfolds, we may experience the sounds as aimless, meandering, arbitrary, lacking direction, and perhaps pointless. Since top-down scene analysis is facilitated by expectation, good part-writing takes advantage of the feeling of leading. When listeners have a sense that a musical line is going somewhere, they have an easier time following it. Said another way, predictability transforms good part-writing into good voice leading.

Let's now apply the above principle to the practice of voice leading, first with regard to schematic norms of pitch successions:

[PR30.] **Follow Tendencies Rule**. *Prefer to resolve tendency tones in the expected (i.e., most common) manner.*

Strong tendencies are most evident in various musical clichés, such as commonplace cadential formulae. However, there exist a huge number of familiar stock musical patterns, most of which are style specific. Examples of such schemata include classical *partimenti*, such as those documented in Robert Gjerdingen's impressive work.[18] Without enumerating all of the possible patterns, are there any simple generalizations that can be distilled from the many existing musical formulae? Traditionally, music theorists have suggested four generalizations.

As we saw in the case of $\sharp\hat{1}$ and $\sharp\hat{4}$, the resolution of a chromatically altered pitch is often dominated by a single continuation tone. As a result, one might offer a derivative of PR30 that focuses on the more particular situation of chromatic pitches:

[PR31.] **Chromatic Resolution Rule**. *When a chromatically altered pitch is introduced, prefer the expected resolution.*

In musical practice, it is not uncommon for a chromatically altered pitch to be preceded by its unaltered form. For example, in the key of C, the pitch G♯ may have been preceded by G♮. In common musical practice, one of the least likely continuations of such a chromatic tone is to reverse direction and return to the unaltered scale tone. Given its rarity, this points to a second, strong prohibition:

[PR32.] **Chromatic Backtracking Rule**. *When a chromatically altered pitch is introduced, don't backtrack by returning to the unaltered pitch.*

Notice that going from G♮ to G♯ involves only a semitone movement. We'd expect such a small movement to occur within a single musical part rather than involving two different parts. If we combine the expectation principle with the pitch proximity principle, then we can infer an even stronger prohibition—the traditional injunction against false relations:

PR33. **False Relation Rule**. *Prefer to avoid two successive sonorities where a chromatically altered pitch appears in one voice but the unaltered pitch appears in the preceding or following sonority in another voice.*

Another way to discourage violations of highly expected patterns might focus on rarely occurring intervals rather than emphasizing uncommon scale-tone successions:

[PR34.] **Avoid Uncommon Intervals Rule**. *Prefer to avoid an infrequently used melodic interval unless you plan to use lots of them (i.e., unless you plan to take advantage of* dynamic-adaptive *expectations).*

So which melodic intervals are uncommon? Most rarely occurring intervals involve large leaps and so are already disparaged by the pitch proximity rule. However, theorists have observed one class of intervals that may be relatively small, yet appear rarely: the augmented intervals. Of course augmented unisons occur from time to time due to tonicizations and modulations (e.g., G followed by G♯), but other augmented intervals are uncommon. Excluding augmented unisons, augmented intervals account for fewer than 1 in 5,000 melodic intervals in chorale harmonizations. The following traditional voice-leading injunction can be regarded as simply a special case of the unusual intervals rule:

PR35. **Augmented Intervals Rule**. *Prefer to avoid augmented melodic intervals, unless you plan to use lots of them (i.e., take advantage of* dynamic-adaptive *expectations).*

It bears emphasizing that these preference rules represent fallible simplifications of the general principle. They are useful mainly for students who have little experience recognizing commonplace musical patterns or schemata—especially recognizing those tones within the schemata that are most constrained in how they normally resolve and so evoke the strongest tendency feelings in listeners.

Compared with PR30, these more specific injunctions have the virtue of providing concrete guidelines. But they derive from the more general underlying principle, best stated in PR30. When it comes to musical expectation, the best advice is: When in Rome (or Naples, or Vienna, or New Orleans …) do as the Romans do. Apart from creating music that conforms to schematic expectations, the perceptual independence of the parts can also be enhanced by helping listeners form accurate dynamic-adaptive expectations. This can be done through repetition:

[PR36.] **Repetitive Patterns Rule**. *Prefer to repeat distinctive melodic patterns, such as themes, motives, figures, subjects, or other sequences.*

Notice that repetition can compensate for passages that violate schematic expectations. Unusual intervals or unorthodox scale-degree successions are less problematic when they are repeated within a work so that listeners come to expect them.

Reprise

Experimental research has led to a distinction between bottom-up and top-down principles in auditory scene analysis. We likened these two approaches to two strategies when assembling a jigsaw puzzle. Bottom-up principles reflect low-level affinities—ways that things seem to fit together. This bottom-up approach is evident in principles like pitch proximity, onset synchrony, source location, etc. Supplementing these heuristics are top-down expectations: we are much more adept at hearing-out some line when we know how it goes. In the language of jigsaw puzzles, we anticipate that a blue puzzle piece is likely to be part of the sky, and therefore located higher in the overall picture.

In this chapter I have proposed a distinction between part-writing and voice leading. Part-writing is a musical practice whose foundation is the bottom-up principles of auditory scene analysis. Voice leading, I propose, is a musical practice that includes part-writing, but supplements the bottom-up principles by adding expectations of "how sounds go," learned through exposure to some musical environment. On the one hand, our expectations might be informed by familiarity with a particular musical work (i.e., veridical expectations). In addition, our expectations are informed by a lifetime of exposure to certain progressions, pitch successions, and tonal affinities (i.e., schematic expectations). Our schematic and veridical expectations rely on our accumulated exposure to a given musical style or culture. Accordingly, voice leading is strongly wedded to culture.

A third form of expectation arises on-the-fly when listening to a novel musical work or passage. We readily form piece-specific expectations in the course of listening (i.e., dynamic-adaptive expectations). Dyanmic-adaptive expectations are facilitated when the music makes use of recurring figures, themes, or motives. There are excellent reasons then, why polyphonic compositions would gravitate toward imitative forms of organization.

Finally, our feelings of "how sounds go" don't just help us in tracking different auditory streams. These feelings also contribute to a sense of direction, movement, tending, or leading somewhere. When the music fails to attend to expected voice movements, the music is likely to evoke an aimless or meandering quality. Predictability transforms good part-writing into good voice leading.

11   Chordal-Tone Doubling

No discussion of part-writing would be complete without addressing the issue of chordal-tone doubling.

An important distinction can be made between *causality* and *correlation*. Two phenomena are said to be *causally* related when one phenomenon produces the other. I blow air into my flute and a sound comes out; the air flow is the cause of the sound. Two phenomena are said to be *correlated* when changes in one phenomenon are *associated* with changes in the other phenomenon. There is a strong correlation, for example, between consumption of ice cream and death by drowning. When ice cream consumption increases, more people drown, and when drowning increases, people consume more ice cream. Now there doesn't appear to be any causal connection between ice cream and drowning. Perhaps people respond to reports of drowning by eating a consoling ice cream cone, or perhaps eating lots of ice cream will cause cramps, which, if one is swimming, might lead to drowning—but these ideas are far-fetched. There is a perfectly reasonable explanation for the correlation between ice cream consumption and drowning: hot summer days tend to encourage people to cool off, and so engage in activities like eating ice cream and swimming. Although ice cream consumption is correlated with drowning, ice cream consumption does not *cause* drowning (nor does drowning cause people to gorge on ice cream).

In research, it is much easier to determine whether two phenomena are correlated than to determine whether one phenomenon causes another. In many areas of research, scholars are unable to figure out what is causing what. In this chapter, we'll see that this is the case for a contentious subset of the traditional part-writing rules.

Most music students learn only one set of part-writing rules. Most of us learn from a single teacher or study from a single textbook. It is only if we encounter someone who studied with a different teacher or used a different textbook that we might realize that there is some disagreement about what rules to follow. In the part-writing canon, the greatest disagreements are to be found in advice about chordal-tone doubling.

When a composer is writing in four parts, one member of a triad will need to be duplicated. For example, when writing a four-note C major chord, either the C, the E, or the G will need to be repeated. But which pitch class should we double? Music theorists have offered lots of recommendations over the years—for example:

Don't double the leading tone.
Prefer doubling the root of the chord.
Prefer doubling the bass.
Prefer doubling the soprano.
Avoid doubling chromatic tones.
Avoid doubling the third of the chord.
Prefer doubling the tonic pitch; next, prefer doubling the dominant.
Prefer doubling the tonic, dominant, and subdominant pitches.
Don't double the third of the chord unless the chord is in first inversion.
For second inversion chords, prefer doubling the fifth.

This list is only a small sample. Over the centuries, lots of rules, heuristics, preferences, and exceptions have been proposed by various writers. Bret Aarden browsed through a sample of pedagogical works and compiled a lengthy list of such rules and recommendations.[1] As with many differences of opinion, musicians often exhibit strong allegiances to certain approaches. As a student, I witnessed animated arguments between teachers about the best approach to chordal-tone doubling.

Broadly speaking, two schools of thought can be distinguished regarding chordal-tone doubling. The schools might be dubbed the *triad member school* and *scale degree school*. The triad member school focuses on advice relating to the doubling of either the root, third, or fifth of a triad. These rules are often qualified depending on the inversion of the chord or the context in which the chord appears. The scale degree school focuses on advice relating to the doubling of particular scale degrees. Some scale tones (like the tonic) are regarded as more suitable for doubling than other scale

Chordal-Tone Doubling

tones (like the leading tone). Once again, these rules are often qualified depending on the chord inversion and harmonic context. The rules may also differ depending on whether the music is in a major or a minor key.

In such matters, an important question is: How do we evaluate the proposed recommendations? How would we ever tell if one set of suggestions is better than another? One approach is to examine the works of composers we regard as especially talented. We may never know the best approach to chordal-tone doubling. But as musicians, we might be happy to emulate the doubling practices of our favorite composers. In short, a good initial strategy is to ask, What do composers actually do?

Working at the Ohio State University, Bret Aarden and Paul von Hippel assembled a database of some 3,600 four-note triads written by Bach, Haydn, and Mozart.[2] They paired each chord with a random four-note triad that was in the same inversion and tessitura as the composed triad. They created the random chords by a computer program without any attention to traditional rules of doubling. Aarden and von Hippel posed the following research question: Given a particular doubling rule, can we use this rule to identify which chords are composed and which are random? Better yet, we could combine rules into various groups and, using a statistical model, test whether one set of rules is better at discriminating the real chords from the random chords.

The initial results are reassuring. When they tested the rules individually, Aarden and von Hippel found that all of the common rules pass muster. The individual rules all perform better than chance in predicting which chords are the real ones. For example, composed triads are more likely to double the root of a major chord in root position. They are also less likely to double the leading tone and less likely to double any chromatic tones.

But there is a problem with interpreting such results. Consider once again the rules related to traffic lights discussed back in chapter 1. Most drivers obey the following rule: *Stop when the light is red and go when the light is green*. As noted earlier, the most common form of color blindness is the inability to distinguish red from green, so color-blind drivers use another rule that takes advantage of the vertical placement of the lights: *Stop when the uppermost light is on and go when the lowest light is on*. Occasionally this rule fails, such as when the traffic lights are arranged horizontally rather than vertically.[3]

In the case of traffic lights, notice that it doesn't much matter which rule you follow. If you are *not* color-blind, you can still rely on the rule: *Stop when the uppermost light is on.* Even though you are following a different rule, your behavior will be identical to the driver who attends only to the color of the lights. Notice also that just because your behavior is consistent with a particular rule doesn't mean that you are following that rule. You might be following a different rule that produces the same (or a similar) result.

Now consider the rule that says *prefer doubling the bass.* For composers like Bach, Haydn, and Mozart, it is indeed the case that the bass is doubled more often than would occur in a random assignment of chord members. But this doesn't mean these composers are following the *prefer doubling the bass* rule. Suppose, for example, that the composers are following another popular voice-leading rule:

Ensure there is no more than an octave separating the soprano and alto; similarly, no more than an octave between the alto and tenor. Any interval can be used between the tenor and bass.

A computer program created to follow this rule will tend to create, purely by chance, chords in which the bass is doubled. When the octave is the largest permissible interval between the soprano and alto, this tends to cause the soprano and alto to be assigned different chord members. The only time the soprano and alto can double the same chord member is when they are separated by an octave (which is the outer limit of separation) or share a unison (which is discouraged). For similar reasons, the alto and tenor will tend to be assigned different chord members. This means that the soprano, alto, and tenor will have a tendency to be assigned different chord members because of the constraints on spacing. The freedom given to the bass has the inevitable consequence that it will be doubled more often than the other voices.

Of course the same problem might plague the rule about spacings between soprano, alto, tenor, and bass. Composers may appear to be following this rule even though this outcome arises for some other reason. Recall from chapter 5 that sonorities tend to be spaced in a way that spreads the partials more evenly across the basilar membrane. When chords are arranged this way, there is less auditory masking. Because critical bands are wider in the bass region, masking is reduced when there are wider intervals separating the bass and tenor, especially when the chord is low in overall register. In other words, the observation that composers tend to double the bass may be purely an artifact of trying to spread partials evenly across the basilar membrane. That is, if a composer aims to minimize masking, then the bass will tend to be doubled.

Notice that this same argument applies to two other common rules: *for second inversion chords, prefer doubling the fifth*; and *don't double the third of the chord unless the chord is in first inversion*. Both rules may arise as artifacts of the rule: *prefer to double the bass*. We now have four sets of rules that all appear to describe the same or similar behavior:

1. For second inversion chords, prefer doubling the fifth. Don't double the third of the chord unless the chord is in first inversion.
2. Prefer to double the bass.
3. Ensure that no more than an octave separates the soprano and alto; similarly, no more than an octave between the alto and tenor. Any interval can be used between the tenor and bass.
4. Arrange concurrent tones so as to minimize auditory masking.

Describing in detail a formal statistical analysis of these rules is beyond the scope of this book, but the results of Aarden and von Hippel's study show that these four sets of rules are essentially equivalent. They all end up producing very similar chord arrangements.

Aarden and von Hippel found many other such conflicting interpretations in their two-year study of chordal-tone doubling. I've only scratched the surface, but you get the idea. Ultimately the most important result from this careful statistical study is that most of the variant rules are redundant. In particular, when pitted against each other, the scale-degree approach and the triad-member approach account for the same patterns in the composed chords. Aarden and von Hippel concluded that scale-degree and triad-member rules are different ways of describing the same musical practices.

In effect, they found something akin to the traffic light scenario. It doesn't really matter which rule drivers follow: the outcomes will be the same. Whether musicians follow the scale degree approach or the triad member approach, the chordal-tone doublings will tend to be the same. A simple comparison helps to highlight this conclusion. Consider two common chord-doubling rules. The scale-degree approach might recommend: *avoid doubling the leading tone*, whereas the triad member approach might recommend: *prefer doubling the root, except in the case of the vii° chord*. What

Aarden and von Hippel found was that the different rules both adequately describe the behaviors of composers.

This could be the end of the story except that some readers might rightly feel uneasy that I haven't identified the underlying causation. What's the *real* motivation in arranging chords? Consider again the rules related to traffic lights. Suppose that a color-blind driver mistakenly ran a red light because the lights were incorrectly positioned. The driver might find a sympathetic judge who would find him or her innocent of blame. But consider now a driver who is *not* color-blind but also relies on the rule related to the position of the lights. When encountering the incorrectly positioned lights, this driver too might end up going through the red light. In this case, the judge would probably not lend a sympathetic ear. We could imagine the conversation:

Judge: You went through the red light?

Driver: The light was in the wrong position.

Judge: But you could see the light was red?

Driver: Yes, but the light was at the bottom of the set of lights.

Judge: Who cares about that. You're guilty!

Here, there is a formally correct rule—the one specified in the traffic code—and that's the rule that specifies *red-stop, green-go*. The position rule works only because it is correlated with the "true" rule. But it's not the "real" rule.

In the case of music, however, there is no formal traffic code. No prior declaration tells us the how and the why. If we knew the true purpose or motivation for the practices of chordal-tone doubling, we could evaluate the efficacy of any given rule or recommendation. However, since we don't know the true purpose, all we can do is compare the rule sets to actual musical practice. Since both sets of rules (the scale-degree approach and the chord-member approach) are equally good at accounting for the compositional behaviors of Bach, Haydn, and Mozart, the only way one could claim that one set of rules is better than the other would be to argue that one set of rules has a better theoretical justification.

So what might the purpose of the traditional rules of chordal-tone doubling be? To my knowledge, no music theorist has argued that these rules serve a unique goal. Of course, there may indeed be one or more unique purposes for these rules that have yet to be identified. But since these rules are traditionally part of the voice-leading canon, this suggests that we entertain the very simple idea that the rules of chordal-tone doubling are intended to serve the same overarching goal shared by all of the other voice-leading rules—namely, promoting the perceptual independence of the individual parts. How then, could chordal-tone doubling contribute to independent musical lines?

Two interpretations of the chordal-tone doubling rules are consistent with the goal of facilitating auditory scene analysis. First, as we saw above, rules like doubling the bass lead to behavior that is suspiciously correlated with efforts to reduce masking. Second, as Aarden and von Hippel point out in their published research, doubling tendency tones can tempt musicians to the transgression of parallel octaves. A preference for doubling the bass reduces auditory masking, and avoiding the doubling of tendency tones reduces pitch co-modulation and tonal fusion. All of these are consistent with the goal of maintaining the perceptual segregation of the musical parts. Although this interpretation might be entirely mistaken, it nevertheless provides a cogent theory that is consistent with compositional practice and also brings chordal-tone doubling within the same explanatory account that unifies the other rules of voice leading.

Following this interpretation, notice that the apparent preference for doubling the bass is an artifact rather than a goal. Doubling the bass is an unintended consequence of spreading out the partials. If this interpretation is correct, then composers don't need to "prefer" doubling the bass. They simply need to be concerned about having larger interval sizes between the lower pitches whenever the sonority is low in overall pitch. In short, there would be no need for a separate rule that encourages musicians to double the bass. If we now accept this theory and assume that the principal purpose of chordal-tone doubling is to enhance the perceptual independence of the parts, then this means that musicians need to consider just one rule. When considering what pitch to double, the sole concern should be to avoid doubling tendency tones. At the end of their exhaustive study, Aarden and von Hippel offer the following pedagogical advice:

If we were to give a summary of doubling rules appropriate for students, we would say this: Worry about the voice leading, not the voice-doubling. Never mind about the triad-member rules, and simply avoid doubling tendency tones. Teachers can justify this by pointing out that doubled tendency tones lead to parallel octaves—a practice that students are already trying to avoid.[4]

There may be other factors to consider when doubling chordal tones. But the simple injunction *avoid doubling tendency tones* covers most of what composers do. To the question, "What is a tendency tone?" we can respond with the answer given in chapter 10: a tendency tone is any tone that has a high probability of being followed by some other tone. More precisely, a tendency tone is any tone that, in some musical context, enculturated listeners strongly expect to be followed by some other tone. Notable examples include chromatic tones and the leading tone, but it also applies to other stock situations, such as the tonic pitch resolving to the leading tone in a cadential 6–4 chord.

When Aarden and von Hippel completed their study, they knew that some theorists would be skeptical of their simplified advice, so along with their published article, they also created a website (now defunct) that allowed musicians to test their intuitions against their model. The Society for Music Theory–sponsored website presented pairs of chords. One chord was drawn from a work by Bach, Haydn, or Mozart. The other was randomly constructed using the same chord members but with three constraints: it (1) included the same bass pitch as the original chord, (2) avoided doubling tendency tones, and (3) avoided more than an octave between each pair of the upper three voices. Visitors to the website were invited to judge which of the chord pairs was created by Bach, Haydn, or Mozart and which by a computer program. After collecting data on the Web for several months, Aarden and von Hippel looked at the ability of music theorists to correctly distinguish between the actual chords and the chords constructed by a computer using the simplified rules. The results were clear: a formal statistical test showed that the music theorists performed no better than chance in distinguishing the composed chords from the chords generated by the simplified model.

On the basis of this research, we might offer the following advice about chordal-tone doubling:

PR37. **Doubling Rule**. *Prefer to avoid doubling tendency tones, such as chromatic notes, the leading tone, or other tones that in context would evoke strong expectations.*

Reprise

There are many important things in life worth worrying about. There are even some aspects of music that warrant thoughtful consideration from conscientious musicians. If we take the research seriously, chordal-tone doubling is not one of them.

12 Direct Intervals Revisited

Perhaps the most peculiar of the traditional rules of voice leading is the direct octaves rule (also known as hidden octaves or exposed octaves). In chapter 7 we showed how three perceptual principles (harmonic fusion, pitch co-modulation, and pitch proximity) led to the following preference rule:

PR23. **Direct Intervals Rule**. *When approaching unisons, octaves, twelfths, fifths, or fifteenths by similar motion, at least one of the voices should preferably move by step.*

Although PR23 resembles conventional statements of the direct octaves rule found in common textbooks, we noted that several commonly included elements are missing. Consider six recent characterizations by textbook authors:

You may not approach a perfect consonance [i.e., fifth or octave] by similar motion … except when the upper voice moves by step. (Laitz)[1]

Avoid leaps in similar motion to octaves or fifths in the outer voices except in chord repetitions. (Benjamin, Horvit, and Nelson)[2]

Direct (or hidden) fifths or octaves result when the outer parts move in the same direction into a perfect octave or perfect fifth with a leap in the soprano part. (Kostka, Payne, and Almén)[3]

Similar motion into a perfect interval with the top voice leaping produces the incorrect motion known as direct or hidden unison, fifth or octave. (Roig-Francoli)[4]

Direct octaves and fifths: when perfect intervals are approached by similar motion … in common practice style (drawing on species counterpoint) these are avoided in the outer voices, but allowed in the inner voices or in any voice paired with a stepwise soprano line. (Clendinning and Marvin)[5]

Fifths or octaves approached by similar motion are called hidden (or direct) fifths or octaves. The term reflects the old theoretical idea that hidden fifths/octaves conceal actual parallels that would occur if the intervals were filled in. (Aldwell, Schachter, and Cadwallader)[6]

Notice the subtle differences among these various statements of the direct octaves rule. Three of the six textbooks restrict the rule to outer voices only. One text limits the injunction to any voice that is paired with the top-most (soprano) voice. Older textbooks often recognize direct intervals as a possibility between any combination of voices. Three of the six texts specify step-motion in the uppermost part as a positive remedy. Other texts imply that a mitigating step-motion can occur in either voice. One text excuses direct intervals provided the same harmony is repeated.

Working at the Ohio State University, Claire Arthur carried out detailed perceptual experiments inspired by the direct octaves rule.[7] She constructed sonorities varying in number from one to nine simultaneous tones. Most of the sonorities contained between three and five tones. Half of the sonorities contained octave doubling of one of the pitch classes. Some of the octaves involved the highest voice, some the lowest voice, and others only inner voices.

Musician listeners were simply asked to identify the number of tones present in each sonority. Consistent with earlier research, she found that the accuracy declined after three concurrent pitches. In addition, she showed that the presence of octaves has a confounding effect. When an octave is present, listeners are much more likely to underestimate the number of tones present in the chord.

In a separate experiment, Arthur preceded each sonority by single tones ("priming" tones). Some of the primes were close to the highest pitch forming the octave, and other primes were close to the lowest pitch forming the octave. The conjecture is that the presence of the primes should draw attention to the ensuing neighboring pitch. That is, pitch proximity should make it easier for listeners to "hear-out" the neighboring tone in the subsequent sonority.

The situation is illustrated in figure 12.1. A double octave is present between D3 and D5. If harmonic fusion takes place, then the harmonics of D5 will be interpreted by the auditory system as harmonics of D3. That is, D3 will effectively "capture" the D5, and so a single complex tone with a pitch of D3 will be heard. Now consider the effect of a single preceding

Direct Intervals Revisited

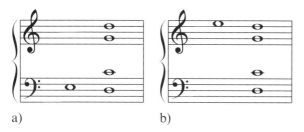

Figure 12.1

neighboring pitch. In figure 12.1a, the bass pitch is preceded by the pitch E3. Due to pitch proximity, we would expect this to cause D3 to be more salient or noticeable for listeners: the pitch D3 will be given a boost. In figure 12.1b, a neighboring pitch (E5) precedes the soprano pitch D5. Once again, due to pitch proximity, we would expect this to cause D5 to be perceived as more salient or noticeable.

Notice that the two cases have different impacts on the potential for harmonic fusion. If we draw attention to the bass pitch D3, there is still a danger that D3 will capture D5 in the soprano since the pitch of the fused tone will be D3. However, if we draw attention to the soprano pitch D5, then harmonic fusion will be less likely. In other words, the benefit of pitch proximity in maintaining stream segregation is greater for the higher of the two voices. This is what Arthur observed in her experiments. In the approach to an octave, step motion is more important for the upper voice than for the lower voice. This result is precisely in accordance with most traditional statements of the direct octaves rule.

However, conventional statements of the direct octaves rule raise another puzzling question. Perceptual experiments show that the highest and lowest voices are more noticeable than other voices in a texture; the highest voice benefits from the high-voice superiority effect, but the lowest voice is also more salient through a mechanism that is not yet fully understood.[8] If inner voices are less noticeable than outer voices, wouldn't we expect the presence of an octave in inner voices to cause greater perceptual confusion than octaves between outer voices? By way of example, consider two sonorities: one contains an octave between the soprano and tenor; the second sonority contains an octave between the alto and tenor. Since the soprano is already the most salient pitch, surely the alto-tenor octave is in greater danger of harmonic fusion than the soprano-tenor octave.

In fact, Arthur's experiments do indeed show that the presence of octaves in inner voices reduces perceptual independence more than when the octave involves an outer voice. Specifically, octaves involving the outermost voices produce smaller errors of density estimation than octaves involving inner voices.[9]

The Muddled Middle

So why do so many traditional statements of the direct octaves rule confine themselves to the relationship between the soprano and bass? One possible explanation begins from the earlier experimental research showing that listeners have trouble tracking more than three concurrent parts. In chapter 8 we identified several reasons why composers might prefer four-part writing even though three-part writing is significantly easier for listeners to follow. If we take the research seriously, it suggests that much or most of the time, listeners do not typically hear four-part writing as consisting of four auditory streams.

If most listeners hear four-part writing as fewer than four streams, how might they be hearing it? In the absence of any pertinent experimental research, we can only speculate. Given the salience of the highest and lowest voices, perhaps most listeners are hearing the inner voices in four-part harmony as a single complex stream. Perhaps most listeners are hearing a *melody + bass + inner-voice accompaniment*. As I emphasize in the next chapter, in dense textures, groups of nominally independent parts frequently amalgamate into a single textural stream.

Notice that if listeners are prone to hear inner voices as forming a single auditory stream, then violations of voice-leading rules involving the inner voices are less likely to cause perceptual confusion. Anecdotally, many common part-writing transgressions do seem less problematic in inner voices than in outer voices. One example is the "frustrated leading tone" discussed in chapter 8. Failing to resolve the leading tone to the tonic seems much more onerous when it involves the soprano or bass rather than the alto or tenor. Similarly, consider the parallel fifths (twelfths) shown in figure 12.2. The parallel fifths involving the alto and tenor (figure 12.2a) sound less vexing than the parallel fifths involving the soprano and bass (figure 12.2b).

If this scenario is correct, it might explain why most traditional statements of the direct octaves rule limit the injunction to outer voices only.

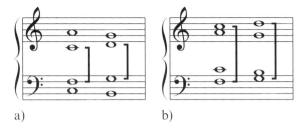

Figure 12.2

Nevertheless, it is unlikely that listeners always hear four parts as three streams: listeners can, and do, hear four streams—just not always, and with difficulty. If our aim is to produce voice leading in which the parts are maximally independent, then it makes sense to remain vigilant against possible harmonic fusion even in cases involving inner voices.

Arthur's experiments appear to vindicate the traditional claim regarding the importance of step motion in the uppermost pitch forming the octave. Drawing attention to the higher pitch reduces the likelihood of harmonic fusion. However, her experiments also show that the presence of octaves (even as static intervals) in inner voices is more detrimental to perceptual independence than octaves involving outer voices. This suggests that it may not be wise to restrict the direct intervals rule to outer voices only. Accordingly, we can go back and revise our original preference rule for direct intervals: we will retain the generality of the rule so that all voice combinations are considered and add that step-motion is advised when approaching the higher rather than lower pitch forming the octave. Hence:

PR23 (revised). **Direct Intervals Rule**. *When approaching unisons, octaves, twelfths, fifths, or fifteenths by similar motion, step motion is preferred preceding the upper pitch forming the interval.*

Reprise

At this point, we might pause and reflect on the component principles that underlie the traditional direct intervals rule. The rule draws on three perceptual phenomena: harmonic fusion, pitch co-modulation, and pitch proximity. The logic behind the traditional direct intervals rule is consequently especially revealing.

When a doctor encounters an obese patient who has a family history of heart disease, the doctor is likely to offer the stern advice, "Don't smoke." (All three factors are implicated in cardiovascular disease, so the combination of the three can prove especially bad.) The logic of the direct intervals rule is similar. In essence, the rule says, "If you are going to violate principle A, and if you are going to violate principle B, for goodness' sake, be sure not to violate principle C as well." That is, if you are going to use an interval prone to harmonic fusion, and you are going to approach it via semblant motion, for goodness' sake maintain close pitch proximity. The traditional direct octaves rule logically implies that all three principles share at least one common goal. The perceptual research suggests that the most likely shared goal is the optimization of stream segregation—that is, the creation of perceptually independent musical lines. Whatever else the direct octaves rule achieves, of all the traditional voice-leading rules, it provides the best evidence we have that the canon of rules has an underlying unity of purpose.

13   Hierarchical Streams

When I was about twelve years old, my father was driving me home from a piano lesson one day when I noticed something I hadn't noticed before: the car horn in our 1965 Plymouth consisted of two pitches, tuned a major third apart. Before, I'd always heard the horn as a single sound, but now I could hear that there were actually two horns. My father (who was quite unmusical) was skeptical of my observation. When we got home, a quick look under the hood confirmed that there were indeed two physical horns connected to the battery. I was delighted. My father must have wondered whether the piano lessons were preparing me for a distinguished career as an auto mechanic.

Over the ensuing months, I became aware that not all cars have two horns. Small cars typically have just one horn producing a single-pitched "beep" sound. I also discovered that the doleful sound of a distant train locomotive was generated by three horns.[1] However, hearing-out these component pitches was not easy. The natural tendency was for the individual pitches to become absorbed into a more cohesive sound object. Hearing-out the constituent pitches required some mental effort.

Similar phenomena happen with musical sounds. A good example of such ambiguous stream perception can be found in the sound of the bagpipes. A single Scottish Highlands bagpipe has four physical sound sources. Three pipes play a constant drone—a bass and two tenor pipes (the latter tuned in unison). The fourth pipe (the "chanter") plays the melody. Oddly, when listening to a single bagpipe player, people typically hear a single instrument playing. If you listen more analytically, you might regard the sound as composed of two lines: the chanter melody plus the static "drone" sound. If you attend even more carefully, you might hear three sounds: the chanter, the bass drone, and the unison tenor drones. Theoretically,

Figure 13.1

Figure 13.2

an attentive listener might be able to hear all four pipes as separate sound sources. So is a bagpipe a single sound source, two sound sources (melody plus drone), three sounds sources (chanter, bass drone, tenor drones), or four sound sources (four physically distinct pipes)? For most listeners, the experience will be of a single source: "a" bagpipe playing. Like the horn in our 1965 Plymouth, many people will tend to perceive a single sound image rather than the constituent sound sources.

Another example of ambiguous streaming is illustrated in figure 13.1. How do listeners hear a sequence of descending thirds? Do they hear a single descending "line of sound"? Or do they hear two distinct parallel descending lines? Does each harmonic third sound like a single sound object—like a car horn? Or are they heard as a sequence of pitch pairs? Many musicians (myself included), would claim to hear both at the same time.

Figure 13.2 shows the opening to the "Dance of the Reed Pipes" from Tchaikovsky's *Nutcracker* ballet. Do listeners hear the triads as three distinct pitches, or do we hear them as a series of chords? In his discussion of this passage, Robert Gjerdingen from Northwestern University draws attention to the tendency for parallel motions to meld into a joint (though perhaps fuzzier) percept; most listeners hear a single rich auditory line rather than three independent lines.[2]

A more unusual example comes from a well-known passage in Stravinsky's *Petrouchka* (figure 13.3) discussed by Emilios Cambouropoulos from Aristotle University of Thessaloniki in Greece.[3] The passage shows a parallel sequence of triads in close position that Cambouropoulos rightly suggests

Figure 13.3

Figure 13.4

will be typically heard as two rich streams rather than six independent lines. In the first five measures, the chords in the upper staff remain at a fixed pitch level while the triads in the lower staff move. The oblique motion between the upper and lower triads suggests two auditory streams. This interpretation is reinforced in the second half of the example where onset asynchrony and contrary motion enhance the independence of the upper and lower staves. But suppose that the triads were not in close position. If each tone in the triads were separated by a tenth from the nearest neighbor, how would the passage sound? Would it be easier to hear six streams? If so, why? Would playing the sequence at a slower tempo make it easier to hear the constituent pitches rather than the chords?

Finally, consider figure 13.4, which shows a simple passage from a Hanon piano exercise. What do listeners make of all the parallel octaves? Do they hear one stream or two parallel streams? Suppose we played just the left hand on an organ with 8' and 4' stops. Would listeners hear two parallel streams an octave apart? To what extent are listeners able to choose what they hear?

To this point in the book, we have emphasized harmonic complex tones as the main objects populating auditory scenes. However, it is clear from these examples that it is also possible to hear complex vertical sonorities as sonic objects in their own right. The reality of chords as perceptual objects is beautifully demonstrated in an extraordinary experiment carried out by Michael Hall and Richard Pastore working at Binghamton University.[4] In the first part of their experiment, Hall and Pastore played pure-tone Es and E♭s and measured the intensity threshold below which each musician listener was unable to detect the presence of these tones. In the second part of the experiment, they repeatedly played a C-G dyad using a piano timbre. They added either a pure-tone E or E♭, and asked their listeners to identify whether the resulting chord was major or minor. During the experiment, they manipulated the intensity of the E or E♭ tones making them progressively quieter. Hall and Pastore made a seemingly astounding discovery: the musicians continued to successfully classify the chords as major or minor even when the sound intensity for the E and E♭ pure tones was reduced below audibility. Some musicians were able to correctly report the chord quality even though the E and E♭ tones were as much as 20 decibels below their hearing thresholds. Said another way, the musicians were able to hear whether a chord was major or minor, even though they were completely unable to hear the crucial third of the chord.

Psychoacoustically, we can infer that there must be some interaction between the three tones that nevertheless remains audible to the listener. Psychologically, we can infer that a chord can be perceived as a basic auditory object without having to hear the constituent parts. In the same way that it is possible to hear a pitched tone without being aware of the resolved constituent partials, it is possible to hear certain chords without being aware of the constituent complex tones. In chapter 15 we'll consider in more detail how this perceptual ability might arise.

So when do listeners focus on perceiving individual pitches, and when do they focus on perceiving emergent chords? Working at the University of Cambridge, Rhodri Cusack, Bob Carlyon, and their colleagues carried out experiments that demonstrate the role of attention in auditory scene analysis. They showed that listeners are sometimes able to direct attention at different levels—what they refer to as "hierarchical decomposition."[5]

Apart from willful attention, the musical context itself can dispose listeners to favor one way of listening over another. When we play a sequence of block chords, one might well imagine that listeners are more likely to favor chord-object perceptions over tone-object perceptions. However, when

Hierarchical Streams

a chord appears in the context of individual pitch lines, then we might imagine that listeners are more likely to favor tone-object perceptions.

The effect of musical context on synthetic or analytic perception is evident in a lovely experiment carried out by music theorist Mark Yeary.[6] Working at the University of Chicago, Yeary played a three-note chord followed by a single probe tone. Listeners were asked to identify whether the probe tone matched the pitch of the middle tone in the target chord. The main manipulation was the context preceding the three-note chord. One context consisted of a four-note melody that would have been expected to stream with the highest note of the target chord. A contrasting context consisted of four three-note chords (none of which contained the final probe tone). Yeary found that listeners abilities to correctly discriminate the presence of the probe tone in the target chord were significantly reduced when the preceding context consisted of chords only—suggesting that a "chordal context" encourages listeners to hear the target chord holistically as a chord-object rather than as three individual tones.

In light of the possibility that listening can occur at different hierarchical levels, it is useful to represent acoustic scenes as hierarchical trees, as illustrated in figure 13.5. The leaves (at the bottom of the tree) represent the resolved partials that constitute each of the sounding tones. The single vertical line at the top of each tree represents the overall synthetic percept—the whole sound experience.[7] In the case of the descending thirds, we hear "a single descending line." However, we may also hear that this descending line is made up of two concurrent complex tones tuned a third apart. In the case of the Tchaikovsky example, the flutes coalesce to form a synthetic

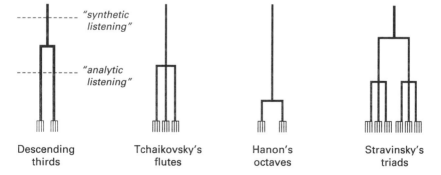

Figure 13.5

melodic line, but a listener may also be cognizant that this line is composed of three flutes.

In these *scene analysis trees*, the vertical dimension represents the degree of analytic attending. Hearing the components at a lower position in the tree requires more *analytic listening*. Conversely, higher vertical positions represent more *synthetic listening*. In synthetic listening, the listener attends more to the sonic forest than the sonic trees. In analytic listening, the listener attends more to the component trees than the forest. Notice how the vertical point of branching differs between the descending thirds, Tchaikovsky's flutes, and Hanon's octaves. The parallel thirds fuse only lightly: it doesn't take much analytic attention to hear-out the constituent tones forming the thirds (hence the higher branching point in the corresponding diagram). Somewhat more listening effort is required in the case of Tchaikovsky's flutes. In Hanon's octaves, even more listener effort may be needed in order to hear-out the two independent parallel lines (hence the lower branching point). The *Petrouchka* example exhibits an extra level in the hierarchy. At the top level, one can hear the overall musical effect; with greater attentiveness, one can hear the "two" triadic parts; theoretically, one could attend to the six individual pitches in each sonority.

When a full orchestra is playing, there are simply too many instruments for any listener to follow. However, even when all of the resources are engaged, there are usually groupings or layers of sound that are more easily deciphered. We may hear a pulsing brass section concurrent with a low bass line, a wash of strings, capped by an overarching woodwind layer. The acoustical scene may consist of one hundred true sound sources, but the result may be the evoking of four broad auditory layers.

Working at Northwestern University, Ben Duane analyzed the hierarchical stream organization in a number of works. Apart from an *auditory stream*, Duane usefully distinguishes between a *musical stream* and a *textural stream*. A musical stream includes the music in its entirety.[8] It is the synthetic percept that represents the singular whole sound experience. A textural stream may be a single instrument, several instruments playing a single part, or a group of instruments or parts that act together.[9]

Having analyzed a sample of musical works, Duane set out to identify what acoustical features are shared in common by the parts forming a *textural stream*. He found that synchronous onsets and offsets were most

Hierarchical Streams

Figure 13.6

important, that semblant motion was moderately important, and that harmonicity has relatively little importance. Unlike auditory streams, textural streams don't necessarily have the same sense of singularity or "oneness." Nevertheless, textural streams are objects of auditory attention; they are sonic "things" to which we can attend.

The concept of a textural stream illustrates that part-writing rules can be used in reverse to integrate sounds, not just to segregate sounds so they are perceptually independent. That is, textural streams help us better understand situations in which apparent transgressions of voice-leading practice remain perceptually coherent. Famously, Claude Debussy made frequent use of parallel fifths and octaves in much of his music. A sample passage from the piano prelude *La cathédrale engloutie* (The Sunken Cathedral) is shown in figure 13.6. Of course, the important point is that these parallel chords are sustained throughout a given passage. The sonorities exhibit high harmonicity, uniform parallel motion, and persistent onset synchrony that encourage them to be perceived as a single textural stream—even if they do not necessarily cohere into a single auditory stream.

The Harmonic Forest

Neurologist Oliver Sacks described the case of Rachael Y., a composer who lost her ability to hear harmony when she was in a car accident that left her with severe head injuries. After recovering from a coma that lasted several days, Rachael discovered that she had lost her perfect pitch ability; more importantly, she had lost her ability to hear chords. All music was now linear or horizontal to her. Major and minor triads were merely three-note collections without any distinctly major or minor qualities. Unable to apprehend harmonic progressions, she could hear only the contrapuntal lines connecting successive pitches.[10]

Before her accident, Rachael composed everything in her head and had good score reading imagery. After her accident, she could no longer perceive or even imagine notated harmonies. She did not lose her intellectual ability to analyze notated chords, but the progressions were purely conceptual. The score could no longer evoke a harmonic experience "any more than a menu can provide a meal," she noted.[11]

Following the accident, harmony no longer existed for Rachael. In effect, she could no longer hear the (chord) forest; instead, she could hear only the constituent (pitch) trees. Moreover, the myriad of pitches produced by a large ensemble now seemed overwhelming. "I absorb everything equally," she complained, "with no filtering system."[12] For Rachael, the passage from Stravinsky's *Petroushka* might evoke six independent lines without any sense of their grouping into two lines each consisting of three-note chords.

Recall that Cusack, Carlyon, and their colleagues provided perceptual evidence for the existence of hierarchical stream perceptions. The phenomenon of *dysharmonia* (harmony deafness) provides additional evidence in support of hierarchical streaming. Apparently brains can wire themselves for experiencing chords or harmonies. In short, textural streams are not merely speculative theoretical constructs. They can be concretely manifested in human brain tissue.

Reprise

As we have seen, an auditory scene can exhibit different levels of organization. At the lowest level, individual partials may amalgamate to form auditory images corresponding to distinct sound sources. At a higher level, these auditory images can combine together to form intermediate musical layers, dubbed *textural streams*. At the highest level, all of the sound sources combine to form a unitary percept of the whole experience, dubbed a *musical stream*.

The experiments by Michael Hall and Richard Pastore, as well as the case of Rachael Y. suggest that attending to different parts of the hierarchy is not exclusively a matter of attention. The ability to focus on different parts of the hierarchy can also depend on an existing skill the listener has for recognizing certain kinds of sonic groupings. Brain damage can destroy a listener's ability to recognize common sonic objects (such as chords), and so restrict how a listener might attend to levels in the scene hierarchy.

The distinction between auditory streams and textural streams means that there is more to part-writing than simply writing parts. There is also the counterpoint of textural streams. In the next chapter, we consider further some of the issues involved in crafting hierarchical musical scenes.

14 Scene Setting

Perhaps the most common musical texture is *tune-and-accompaniment*. Examples of tune-and-accompaniment include a singer with guitar, a solo instrument with piano accompaniment, and a solo piano work in which a melody line is accompanied by, say, an Alberti bass figure. There are many different styles of accompaniment. Figure 14.1 shows scene analysis trees illustrating hierarchies for two tune-and-accompaniment scenes. Recall that the vertical dimension represents the degree of *synthetic* or *analytic* listening—with the most synthetic corresponding to the top of the tree and the most analytic corresponding to its bottom. In both diagrams, the rightmost branch represents a single-stream "tune." Figure 14.1a represents a tune plus strummed guitar accompaniment. Strummed chords may consist of up to six sounded strings; there is no attempt to create six independent lines within the accompaniment, so the guitar tones exhibit strong cohesion—forming unitary chord perceptions. Figure 14.1b illustrates a

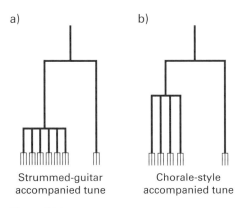

Figure 14.1

tune plus four-part chorale-style accompaniment. Here, part-writing conventions have been followed so that there is greater voice independence in the accompaniment parts; an analytic listener is more likely to hear-out one or more accompaniment lines. Notice that the vertical branching point for the strummed guitar accompaniment is lower than the branching for the chorale-style accompaniment, reflecting the greater difficult in hearing-out the component tones produced by the guitar.

The scene analysis trees in figure 14.2 illustrate a stride bass accompaniment in which a bass note alternates with off-beat chords. In stride bass, the alternation between bass and chord typically involves a wide pitch separation. Like yodeling, recall that this separation is likely to cause perceptual segregation into independent lines. However, also recall that segregation also depends on the speed of pitch alternation (see figure 6.3 in chapter 6). The more rapid the alternation, the more likely it is that the texture will split into two streams. If a stride bass is played at a slow tempo, the effect is more likely to be heard as a single stream, like slow yodeling. At a faster tempo, the stride bass is much more likely to be perceived as an independent bass line accompanied by a concurrent stream of off-beat chords (the boom-chick-boom-chick effect). The effect of tempo is reflected in the

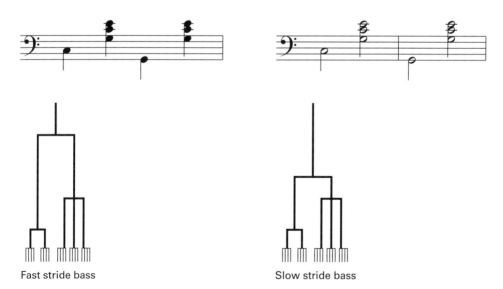

Fast stride bass   Slow stride bass

Figure 14.2

corresponding scene diagrams. The main branch between bass line and off-beat chords is higher in the faster tempo passage reflecting the greater likelihood that the two lines will segregate perceptually.

In the previous chapter, we discussed Tchaikovsky's "Dance of the Reed Pipes." However, the discussion focused exclusively on the three flute parts, ignoring Tchaikovsky's accompaniment. In the complete orchestral version, a stride bass-like accompaniment is provided using pizzicato strings. The revised scene analysis tree is illustrated in figure 14.3. The dotted shadow lines indicate that multiple instruments are playing in unison. The three flutes form a melodic texture stream, while the stride bass forms the accompaniment texture stream. Notice that the branching of the accompaniment texture is higher than for the melodic texture. This reflects the fact that it is much easier to hear-out the lower and upper tones in the stride accompaniment than to hear-out the individual flutes.

Although some musical works maintain a static scene hierarchy throughout, many works include changes of scene. Some of these changes are modest. For example, in "Dance of the Reed Pipes," a bassoon obbligato part enters after several measures, introducing a single additional stream. Other changes may involve dramatic contrasts of texture. The scene analysis trees illustrated here are static snapshots that pertain to a given moment; more accurate portrayals would require animated trees that evolve over time.

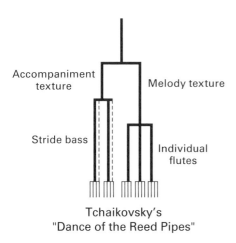

Figure 14.3

Figure 14.4

Changing the scene hierarchy requires care. Consider once again the stride bass accompaniment, this time as notated in figure 14.4. The change here simply involves a decrease in the pace or tempo of the figuration in the second and third measures. As already noted, tempo has a direct effect on stream formation. At rapid tempos, pitch alternations are more likely to segregate into separate streams; at slower tempos, there will be a greater sense of connection between widely separated pitches. Consequently, listeners are more likely to hear two streams in measures 1 and 4, and one stream in measures 2 and 3. Note that there is nothing wrong with the slow version of the stride bass, just as there is nothing wrong with sustained passages involving parallel octaves. The concern is with the stability of the scene hierarchy and the ability of listeners to parse the scene changes with minimal confusion.

Acousmatic Scenes

No development offers more creative potential to manipulate the listener's experience of an acoustic scene than the advent of computer technology. A compelling illustration of the musical manipulation of auditory streams can be found in *Dreamsong* (1978) by the American composer Michael McNabb.[1] *Dreamsong* is an example of "acousmatic" music—music that is composed and exists only in electronic form. The work begins with a cacophony of speech—like a crowd of voices you would hear in a reverberant lobby during a concert intermission. At the same time, electronic bells "toll" in the background. After a few seconds, the bells disappear and the lobby sound is treated in the manner of a loudspeaker being swung on a rope. You can hear the phasing of the sound increase, moving like the swoosh of a passing car. The swoosh evolves into a pitched electronic-sounding tone that traces a slow upward arpeggio until stabilizing on a single pitch. The tone gradually morphs into the sound of a female vocalist singing "ah." This single voice becomes a unison chorus of many singers, returning back to a single voice, and then reassembled into a complex vocal chord. The chord returns to an electronic-sounding timbre, and a high-pass filtering effect is used to draw attention to specific harmonics rising up in sequence.

The piece continues in this fashion for the full nine-minute duration of the work. It is a masterful display of evolving acoustic (and therefore) auditory scenes: tones morph into chords, separate pitches converge into single complex sounds, noises amalgamate into tones, and tones blossom into noises. The work ends by returning to the original lobby sound that began the work; however, the lobby sound itself begins to speak as though a single voice, finally morphing into the recorded sound of Dylan Thomas reading the final lines of his poem, "Love in the Asylum": "And taken by light in her arms at long and dear last, I may without fail suffer the first vision that set fire to the stars."

Dreamsong is but one example of an entire genre of music in which recorded natural sounds are digitally processed, resynthesized, and blended into a continuously evolving soundscape. Digital audio techniques of the type featured in McNabb's work allow composers access to the full range of hierarchical acoustical scene treatments glimpsed in our more conventional examples. Compared with the refined control available through digital processing, Ravel's orchestration in *Bolero*, Tchaikovsky's treatment of three flutes in the *Nutcracker*, and Stravinsky's writing two-part counterpoint with triads might seem primitive. The important point is that composers have been creating these transformative effects for centuries. Electronic media simply make the techniques more accessible and, in many cases, more robust.

With electronic manipulation, the musician effectively possesses a sort of zoom lens that allows the listening experience to range between analytic and synthetic perceptions. Even with a seemingly static scene, the composer can conduct a dynamic tour through the scene hierarchy. Zooming in, an orchestra is broken into sections, sections into individual instruments, which in turn disintegrate into partials. Zooming out, an entire orchestral texture is compacted into the quiet tinkling of an ear bud—which in turn becomes one of hundreds in an electronics shop, and where the whole shop becomes just one source in the cacophony of a city soundscape.

In creating these effects, the pertinence of voice leading to the organization of spectral and acousmatic music has not gone unnoticed.[2] Dutch musicologist Leigh Landy has rightly spoken of the concurrent layers of sound as "a modern equivalent of counterpoint."[3] Landy is not the only musician to have noticed the relationship.

Designing Musical Scenes

One of the main compositional tasks in any musical style is scene setting. A composer sets a stage, populates the acoustical scene with one or more sounds, and choreographs their entries, mergers, contrasts, evolutions, fragmentations, and final exits. In conventional music-making, the building blocks are instruments and voices. In electroacoustic music-making, the building blocks can also include individual partials. In either case, complex, dynamically changing scene hierarchies can be created.

When composing a musical texture or acoustic scene, the musician faces the same questions at each moment in time: How do I want listeners to experience this? That is, what auditory scene do I want listeners to parse from the acoustic scene I'm creating? How many concurrent streams do I want listeners to hear? Do I want the listener to hear each stream as equally important, or do I want particular streams in the foreground or background? How do I ensure that the components of this auditory or textural stream remain fused? How do I ensure that a given stream is not inadvertently absorbed by another? Have I created too many independent streams for listeners to follow? How should the auditory scene evolve over time?

Problems arise not when streams fuse or fork, but when listeners become momentarily confused. The parallel octaves in Debussy's *Sunken Cathedral* cause no problem because they form a single coherent texture stream. But a single parallel octave in the midst of two-part polyphony causes momentary "dropout" where listeners can experience a "now you hear it, now you don't" discontinuity in the auditory scene. An abrupt change of tempo in a stride-bass accompaniment produces an analogous effect to the occurrence of parallel octaves. Both the parallel octave and the stride-bass tempo change produce a temporary "dropout" in the scene hierarchy. Both the octave and the tempo change invite listener confusion when parsing the auditory scene. Changing from one musical texture to another requires care. Part of what makes McNabb's *Dreamsong* so compelling is that the transitions between difference scene hierarchies are so seamless.

It should now be clear that the principles of auditory scene analysis provide the enabling tools by which these musical scenes or textures are interpreted by listeners. It makes sense then, that these same tools would play a prominent role in designing the musical scenes in the first place.

When considering the voice-leading canon, the natural tendency is to regard the rules as pertaining to the creation of perceptually independent parts or voices. But also implied in the underlying principles is the reverse process by which musical instruments, voices, or parts can be integrated to form cohesive perceptual lines. The rules tell us how to make musical parts perceptually independent, and simultaneously give us tools for making musical parts cohere into textural streams.

If you want a set of disparate instruments to cohere as a single textural stream, the best results will arise when they employ synchronous onsets, move in parallel, play harmonically related pitches, are positioned close together in space, use homogeneous timbres, include more than three instruments, and so on.

In other words, voice leading provides the toolkit for designing auditory scenes. The rules of voice leading aren't simply tools for creating polyphonic music or Baroque-style chorales; they are tools that allow composers to construct and control any kind of musical texture. Voice leading truly is the art of combining concurrent musical lines.

Why the Baroque Canon?

Music lovers today have access to a prodigious range of musical styles, genres, and cultures. Most of the music we listen to (and enjoy) does not consist of four-part chorale-style harmony. So why do nearly all introductory music theory textbooks focus on Baroque voice-leading rules and chorale-style SATB harmony to the virtual exclusion of other types of part-writing?

Before we address this question, let's first spend a little time considering some alternatives. Traditionally, it has been common to distinguish three broad practices in Western part-writing: those associated with the Renaissance, Baroque, and Classical/Romantic periods. However, this three-part division represents only the broadest brush strokes. There are innumerable differences in specific genres and even in the part-writing practices of individual composers. Knowledgeable historians of theory readily point to differences between *cantus firmus* counterpoint, Fuxian species counterpoint, School of Notre Dame polyphony, Lassus-style motets, English madrigal style, late Baroque fugue style, Lutheran chorales, Bartók's strict imitative counterpoint, and many more. Apart from these more polyphonic textures, there are also innumerable other musical textures, such as close harmony arrangements, Kentucky folk organum, South African hymn singing—the list is long.

Some of these part-writing practices were described and formalized in contemporary treatises that prescribed specific compositional rules or recommendations. Other historical practices have been formally described only by modern scholars, and there is little evidence that composers in the past were aware of the "rules" they were following. The number of proposed rules and recommendations in both historical and modern accounts is mindboggling. For example, in his classic study of Renaissance modal counterpoint, McGill University scholar Peter Schubert distills the complex writings of a dozen or so theorists who wrote in a relatively brief period (from 1558 to 1622). Schubert manages to condense this practice into just over seventy rules, often simplifying and clarifying some rather complicated injunctions. As just one example, Schubert describes what he calls the *pyramid rule*: when using skips and steps in the same direction, the larger interval should be lower than the smaller one. He offers the analogy of climbing stairs: "You are more likely to take stairs two at a time when you start going up, switching to one at a time near the top."[4] In short, place the bigger interval between the lower pitches in a musical line.

Sure enough, when I carried out statistical analyses of choral music, I found that the pyramid rule is evident in works throughout the Renaissance. In Baroque choral music, the pyramid rule is still evident, but not for the bass voice. By the Classical period, the pyramid rule is evident only in the soprano voice.[5] In other words, Schubert's pyramid rule seems to describe a particular part-writing practice that flourished during the Renaissance but slowly disappeared.

Differences between historical periods are often quite marked. In Renaissance species counterpoint, all perfect intervals must be approached by contrary or oblique motion. By the Baroque period, approach by similar motion was also permitted as long as one of the parts moved by step (although many writers added the provision that the step motion must be in the upper-most voice). Baroque practices, however, were far less lenient about part-crossing compared with Renaissance practices.

These examples represent only a handful of the innumerable differences evident in various periods and genres. Sustained scholarship over the past half-century has made music educators much more aware of the peculiar narrowness of Baroque four-part chorale-style voice leading. When asked by students why they should learn this particular practice, educators have had to resort to a remarkably weak argument: "It is just one culturally and historically situated musical practice, that is neither better nor worse than other practices. Through the inertia of tradition it has remained part of the core curriculum. It's probably good for students to gain some experience with it."[6]

If this rationale were indeed correct, then something is truly amiss if virtually every musical institution continues to insist on teaching Baroque chorale-style practice as a core subject. It looks like some strange conspiracy.

The research reviewed in this book offers a more cogent explanation for the continued popularity of Baroque chorale-style training. No other historical practice conforms so closely to modern perceptual and cognitive research regarding how independent sounds are heard in complex acoustic scenes. That's probably not a coincidence: the Baroque canon of voice-leading practices congealed around the time when polyphonic music-making reached perhaps its greatest popularity—and arguably witnessed its most accomplished practitioners.

Of course, the correspondence between late Baroque practice and modern auditory research is not perfect. Notably, the conventional voice-leading canon does not include several factors that modern research recognizes as important. Of the perceptual principles known to affect auditory scene analysis, SATB chorale-style part-writing excludes most of the easy segregation techniques. The use of different timbres is very effective in allowing listeners to correctly segregate independent sound sources, yet chorale-style writing largely assumes homogeneous instrumentation. Sound sources are easier to segregate when they provide unique localization cues, yet chorale-style writing doesn't assume that soprano, alto, tenor, and bass voices are widely dispersed in space. Asynchronous onsets are powerful ways of segregating sources, yet chorale-style writing emphasizes block chord progressions over a more polyphonic rhythmic texture. Limiting the textural density to just two or three streams is sure to facilitate stream segregation, yet chorale-style writing dictates a four-part texture—a density that just exceeds the common limit for easy scene parsing. Moreover, those four parts are explicitly limited to a total range of just three octaves, which limits the opportunities for wide pitch spreads and so adds to potential conflicts arising from pitch proximity.

By excluding the use of these principles, chorale-style part-writing forces musicians to rely on only a few segregation principles, notably pitch proximity, pitch co-modulation, and harmonicity. Notice that mastery of chorale-style part-writing effectively teaches musicians how to deal with the thorniest of stream segregation problems: those involving homogeneous timbres, restricted pitch range, synchronous onsets, and high-density scenes. A musician who has mastered these principles will have little difficulty arranging coherent musical scenes when allowed to make use of other tools in the scene analysis toolkit, such as varying instrumentation or using asynchronous rhythms.

Reprise

Any musical texture amounts to a sort of theatrical stage occupied by one or more sound sources or activities. Musicians literally set the scene, and auditory scene analysis is the process by which listeners subjectively apprehend that scene. Scenes are typically hierarchically organized with individual partials combining to form images, auditory streams, then textural streams, and ultimately overarching musical streams.

Scene analysis trees provide useful ways of visualizing an acoustic scene; they also allow us to consider how that scene might be parsed by listeners into a corresponding auditory scene. Since many musical works exhibit a dynamic or evolving scene, scene trees merely represent snapshots of particular musical moments. Nevertheless, scene trees can be helpful in analyzing the sonic experience. In particular, changes over time can alert us to possible parsing confusions for listeners—such as when a stream momentarily drops out.

How listeners parse acoustic scenes is described by the principles of auditory scene analysis. For composers who care about such matters, these same principles provide guidelines for how to create coherent musical scenes. The traditional rules of voice leading clearly echo the principles of auditory scene analysis. However, rather than arising from formal scientific experiments, these rules arose by intuition distilled from centuries of musical practice. In this chapter we have noted that the voice-leading canon is not simply about ensuring the perceptual independence of concurrent lines of sound. When the same principles are applied in reverse, they help to ensure perceptual integration—where seemingly disparate elements combine to form a single line of sound. In other words, the canon provides a useful compositional toolkit for creating a huge range of possible scene hierarchies, from yodeling, to tune-and-accompaniment, to complex electroacoustic textures.

Why do introductory music theory textbooks focus on Baroque voice-leading rules to the virtual exclusion of other types of part-writing? The evidence suggests that late Baroque practice most closely reflects known principles of auditory scene analysis. That is, late Baroque polyphonic practice represents the clearest historical statement of how to make perceptually independent lines of sound—unencumbered by other possible musical goals.

15 The Cultural Connection

What is a sound? Philosophers will recognize this question as a specialized form of the more general question: What is a thing? As we saw in chapter 13, a "thing" depends on the scope of our gaze. A "sound" like a car horn might entail more than one physical sound source, and each source might exhibit many partials. Similarly, "sounds" might be amalgamated together to form larger objects like chords. Once again, what constitutes a sonic object seems to depend on the scope of our gaze.

In the previous chapters, we have discussed bottom-up and top-down principles of auditory scene analysis. The tendency is to think that top-down principles are learned, whereas bottom-up principles are innate. For example, in chapter 4, we explored the *harmonic sieve* concept—the idea that the auditory system has some sort of template for recognizing partials that form a harmonic series. Although I didn't claim so, this principle looks like it might be prewired—part of the standard equipment for human brains.

When discussing biology, it is common for two concepts, *innate* and *universal*, to get confused. Something is innate when it is biologically preordained, such as when a particular gene leads to blue eye color. Something is universal when everyone has it (such as having a nose). Something can be innate but not universal (such as blue eyes). It is also possible for something to be universal but not innate. For example, nearly every human culture engages in fire making, but no biologist has proposed that there is a fire-making gene.[1] Similarly, all cultures form a belief that "the sun will rise tomorrow," but no one posits a "sun-will-rise-tomorrow-belief" gene. There are universal phenomena that arise through learning and innate phenomenon that are not universal.

Although the harmonic sieve appears to be a universal property of human audition, it may not be innate. In fact, Ernst Terhardt, the father of psychoacoustic research on pitch, famously argued that the harmonic template is learned from simple exposure to harmonic sounds. Since harmonic sounds (like the human voice) are found everywhere in the world, all humans with intact hearing will develop a harmonic template and the corresponding perception of pitch. According to Terhardt, sensitivity to harmonic sounds is universal but not innate.

Over the past two decades, Annemarie Seither-Preisler and her colleagues at the University of Graz in Austria have assembled a large body of experimental research consistent with Terhardt's conjecture that the perception of pitch depends on learning.[2] Her work has demonstrated that familiarity with different sorts of sounds can cause changes in how an individual hears pitch. Her research group has shown how different sound exposures cause plastic changes in the human auditory cortex.[3] She has also documented differences in pitch perception between musicians and nonmusicians.[4] Corroborating research has been carried out by Peter Schneider and his collaborators at the Neurological Clinic of Heidelberg University Hospital. Psychoacousticians have identified two extremes of pitch perception. At one extreme are listeners who attend mostly to the periodicity of the fundamental frequency for a complex tone. At the other extreme are listeners who attend mostly to the spectral envelope. (The details are not of concern here.) What is interesting is that these two different approaches to pitch perception emphasize different parts of the brain. *Fundamental pitch listeners* favor pitch processing on the left side of the brain, whereas *spectral pitch listeners* favor pitch processing on the right side of the brain. In 2005, Schneider and his colleagues measured the degree to which 409 professional and amateur musicians favor fundamental versus spectral pitch processing. The results are reproduced in figure 15.1, grouped according to instrument played.[5] As can be seen, the preferred method for pitch processing is closely related to the musician's instrument or voice. In general, instrumentalists who play percussive instruments are more likely to favor fundamental pitch processing. Instrumentalists who play low-pitched instruments that tend to produce mostly sustained tones are more likely to favor spectral pitch processing. This includes the bassoon, cello, horn, saxophone, and organ. Higher-pitched instruments tend to shift processing toward the fundamental-pitch-processing end of the spectrum. Although people might

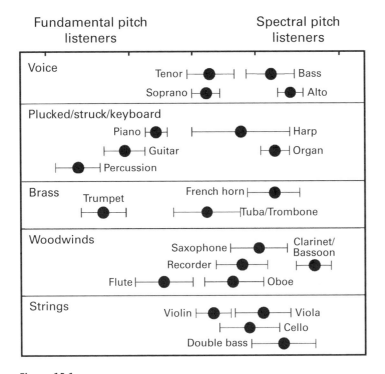

Figure 15.1
The degree to which different musicians favor *fundamental or spectral* brain processes when hearing the pitch of any sound. The study involved over 400 musicians; results are grouped according to instrument played or voice. Bars indicate standard error rather than range. Results suggest that the sounds a person hears most often change the way the person perceives sounds in general. *Source*: Schneider, Sluming, Roberts, Bleeck, and Rupp (2005). Reprinted by kind permission.

choose to play an instrument based on their existing pitch-processing preference, the causality is likely the other way around. For example, singers don't generally get to choose whether to sing bass or soprano. The evidence suggests that the sounds a person hears most often change the way he or she processes sounds in general.

Perhaps the most dramatic evidence concerning the learning of pitch is to be found in the experience of cochlear implants. Some kinds of deafness can be alleviated to some degree through the use of implant technology. Cochlear implants consist of a set of very thin wires that are surgically inserted into the cochlea. The bundled wires spiral around the cochlea, each wire making contact with a different point along the basilar membrane.

The patient wears a microphone attached to a signal processor that filters the sound into a series of frequency bands or channels. These channels then stimulate the corresponding place along the basilar membrane. Of the roughly 3,500 sensory cells found in a healthy cochlea, cochlear implants typically reduce the resolution to the equivalent of just 16. Although the basilar membrane itself is no longer responsive to sound, the electrical stimulation of the sensory neurons allows the patient to hear something.

For several weeks after the implant has been inserted, the patient typically hears very distorted and unintelligible noise. But over time, the auditory system "wires" itself, and a coherent world of sound slowly emerges. For surgical reasons, the implant is not inserted all the way to the apex. Since low frequencies are activated toward the apical end, cochlear implants fail to convey any low-frequency information. In 2005, cochlear implant patient Michael Chorost published a book describing his experience of having lost his hearing entirely and then the process of regaining his hearing through an implant. Chorost describes the rapid speed by which he regained the pitches of sounds. Initially the cochlear implant made his own voice sound like he was a soprano. However, his adaptation was swift: "My own voice sounded low-pitched to me again within a few hours. By the next day, I could differentiate between male and female voices; by the next, male voices sounded deep; by the next, women sounded like women again. My brain had somehow reinterpreted a huge frequency change back into a semblance of normality."[6] Not all implant patients experience such a rapid or dramatic adaptation, but Chorost's experience highlights the extraordinary plasticity of the auditory system. It also reinforces our earlier lesson that pitch is a subjective auditory phenomenon rather than an "out-there" acoustical phenomenon.

The Sound of Speech

Learning is most apparent when environments differ. Of all the sounds we experience in the world, perhaps the most profound differences between one environment and another are to be found in language.

Depending on your culture, you will be exposed to different speech sounds. For example, a small number of the world's languages employ clicks as part of the basic inventory of speech sounds or phonemes. Well-known click languages include Nama, Sotho, and Xhosa. In the same way that many languages employ "plosives" like *t*, *k*, *b*, and *p*, these languages also make use of clicks formed by pressing your tongue against the roof of your mouth and rapidly pulling the tongue down. Depending on the placement of your tongue, different click sounds will be produced. The Nama language uses over a dozen different click sounds.

To people who speak languages without clicks, listening to a language like Nama is an odd experience. (Sound examples can be heard on the Web.) To non-Nama ears, it sounds like a person speaking a foreign language while someone else is making random clicking sounds. To our ears, it sounds like two auditory streams: a speech stream and an independent stream of clicks. Of course, Nama speakers don't hear the sounds at all this way. They hear the clicks as an integral part of a single speech stream. How do we know this? If they weren't able to hear the clicks as connected to the other speech sounds, then they wouldn't be able to understand what the person is saying. In the same way that we hear the *t* as connected to the *oo* in the word *too*, Nama speakers hear the clicks as connected to the surrounding phonemes. How the sounds stream depends on what language you speak.

Another useful demonstration can be found in an unusual phenomenon known as *sine-wave speech*. (Sound examples can be found on the Web.) In ordinary speech, each vowel is distinguished by a particular set of resonances called "formants." For example, if a sound emphasizes partials in the vicinity of 300 Hz, 2,000 Hz, and 3,100 Hz, it will have an *ee*-like quality. If a sound emphasizes partials near 200 Hz, 900 Hz, and 2,000 Hz, it will have an *ah*-like quality. Remember that these are just broad resonant regions. Any partials near these frequencies will be boosted in energy. Typically several harmonics are emphasized for each formant.

In the early 1980s, Robert Remez and Philip Rubin at the Haskins Laboratory in Connecticut tried to create synthetic speech by replacing each resonance by a single frequency. A normal spoken vowel will typically produce twenty or thirty significant harmonics, and each resonant formant will be associated with several prominent harmonics. However, Remez and Rubin distilled the entire vowel down to just three frequencies—one frequency representing each of the three main formants. They went on to create synthetic sentences by having the three frequencies glide around as though they were formant resonances in a speaking voice.[7]

When you first hear sine-wave speech, it sounds something like a flock of twittering birds. It doesn't sound like speech at all. As you continue to listen to it, however, you find that it begins to take on speech-like qualities. After some further exposure, most listeners start to understand what the "twittering birds" are saying. (I highly recommend the Web examples.)

You might suppose that listeners will switch back and forth between hearing either the twittering birds or the speech. Many perceptual experiences are either/or phenomena, although the experience may flip-flop between two different perceptions. You perceive either one or the other, but not both at the same time. A classic example is the famous faces-vase illusion devised by the Danish psychologist Edgar Rubin (no relation to Philip Rubin): viewers see either a vase or two facial profiles but not both concurrently.[8] In the perception of auditory streams, there is a similar bistable tendency to hear either one or another arrangement.[9] However, in the case of sine-wave speech, experienced listeners report being able to hear both the twittering birds and the speech at the same time.

In the case of the click languages, whether we hear the sounds as forming one stream or two streams depends on our cultural background. In the case of the sine-wave speech, whether we hear several concurrent gliding streams or a single integrated speech stream (or both at the same time) depends on our familiarity with sine-wave speech and where we place our attention. Sometimes listeners hear the acoustical forest (where all of the partials amalgamate into speech sounds) and sometimes listeners hear the individual acoustical trees (where the partials segregate into several twittering birds). As already noted, with sine-wave speech, listeners often report being able to hear both the forest and the trees at the same time. The experience is similar to hearing pitches and chords at the same time.

Language and Closure

The influence of language on auditory perception is beautifully illustrated by research on rhythmic grouping. In 1909, psychologist Herbert Woodrow (a cousin of President Woodrow Wilson) published an influential study concerning the grouping of tones. Woodrow found that a *short-long* tone sequence (e.g., dit-dah) is more likely to be perceived as complete or closed compared with a *long-short* tone sequence (dah-dit).[10] That is, long tones are more likely to signal the end of a rhythmic group. Woodrow's work was replicated several times, and for the past century, it was assumed that "group final lengthening" was a cross-cultural universal—that is, until a study by John Iversen, Aniruddh Patel, and Kengo Ohgushi showed that Japanese listeners have the opposite experience. For Japanese listeners, a long-short tone sequence tends to sound more complete or closed than a short-long sequence.[11]

These perceptual differences echo an important difference between the English and Japanese languages. Most European languages employ so-called *proclitics*—brief monosyllabic words (like *the* and *were*) that precede content words. As a result, word groups tend to form weak-strong or short-long clusters like "the car," "to work," and "were gone." By contrast, Japanese employs *enclitics* (unstressed suffixes like *-ga* and *-desu*) that bind to the ends of content words. Consequently, the meaning cluster tends to form a strong-weak or long-short pattern. Speakers of most European languages (including English) typically learn to parse short-long as group terminating, whereas Japanese speakers learn to parse long-short as group terminating. It is not yet known whether these differences are also evident in music perception.[12] Western music does indeed exhibit group-final lengthening where the last tone in a rhythmic group or phrase tends to be longer than the preceding tones. However, we don't yet know whether native Japanese speakers hear the Western "ta-dah" cadence as sounding slightly less closed or final than is the case for those whose first language is drawn from Europe.

A Lesson

As we have seen, how musicians perceive pitch depends to a considerable degree on the instruments they play. How we hear rhythmic grouping depends to some extent on the language we speak. In addition, our language facility influences whether we hear clicks as part of speech or as an independent auditory stream. What all of these examples imply is that the sound environments we are exposed to shape how we hear. Our auditory systems wire themselves in response to the kinds of sounds we experience. In the case of cochlear implants and sine-wave speech, the auditory system appears to be quite plastic. There is considerable scope for learning (and relearning) how to perceive the world. When we attempt to parse an acoustic scene, much of the auditory skill we bring to the task has been conditioned by experience. Apparently it is not just top-down aspects of auditory scene analysis that arise from learning. At least some of the bottom-up templates we use also arise from learning. Since many templates can coexist in the auditory system, it may be possible to hear a given acoustic scene in more than one way. We may hear chords *and* the constituent pitches. We may hear "a bagpipe" and simultaneously hear "drone plus melody." We may hear "an octave" and also "two pitches."

So why, at the age of twelve, was I able to decipher that the horn in our 1965 Plymouth consisted of two pitches while my father could hear it only as a single sound? Didn't we both live in the same acoustical world? If scene analysis adapts to the sonic environment, why wouldn't my father and I hear the same sound objects? Of course, we didn't live in exactly the same acoustic environment. Unlike my father, my childhood involved lots of music lessons and many thousands of hours of practice. What was it specifically about my piano lessons that led me to be able to hear the car horn in two different ways? By playing the piano, I was exposed to lots of different combinations and permutations of pitched sounds. I might play the pitch C, and the pitch E, and both C and E together. Over the years, my auditory system was exposed to roughly a million instances of harmonic major thirds and a comparable number of melodic major thirds. Moreover, my exposure was not simply passive perception. My perception was linked to various motor activities. I was "acting out" individual tones by moving individual fingers and "acting out" chords by moving an entire hand. Some sound objects were being created "by the handful," whereas other objects were created one digit at a time.

Even though we lived in the same house, my father and I nevertheless lived in subtly different sonic worlds. I had been exposed to far more combinations of thirds—in both harmonic and melodic contexts. I simply recognized the major third produced by the car horn, inferred that there must be two constituent pitches, and was able to hear-out the component tones. From an implicit learning perspective, there were surely enough differences in auditory exposure for my father and me to perceive the same sound in different ways.

Reprise

In chapter 10, we saw how top-down expectations influence stream perceptions and noted that these expectations are learned from experience.

In fact, most auditory researchers believe that learning is the sole source for top-down auditory scene analysis. In this chapter, we've explored some evidence suggesting that some of the bottom-up principles may also be learned—or at least are susceptible to environmental influence.[13]

The point of this chapter has not been to establish that auditory scene analysis is entirely a learned phenomenon. Instead, my more modest aim has been to dampen or discourage any assumption that the brain mechanisms used to achieve auditory scene analysis are innate and immutable. While the disposition to analyze auditory scenes is surely an evolved innate behavior, the means by which this is achieved likely involves a mix of innate and learned mechanisms.

As we've seen, the sonic environment plays a formative role in various aspects of auditory processing. This means that cultural background and individual experience may be directly relevant to our understanding of voice leading. To the extent that auditory scene analysis principles are conditioned by the sound environment, the implication is that voice-leading principles are also conditioned by the musical environments a listener experiences. It would be wrong to assume that everyone parses an acoustic scene in the same way.

We don't yet know the extent of the auditory system's malleability. However, the existing research suggests that a full understanding of voice leading may not be achieved without attention to the specific histories of various musical cultures, subcultures, and styles.

16   Ear Teasers

After the passage of some four centuries, composers today still rely on traditional part-writing rules when creating at least some of their music. Why have so many composers done this? What is gained by clarifying the auditory scene? In particular, why is multipart music compelling to musicians and listeners alike? What is the lingering appeal of this nominally Baroque practice?

Many reasons could be given, but the most common one offered by composers (and teachers) is that part-writing "sounds better" when the composer follows traditional voice-leading practice. This claim suggests that at the heart of voice leading lies an aesthetic question: Why does proper voice leading sound better?

Like all other sensory systems, the purpose of the auditory system is to provide listeners with relevant information about the world. As we saw in chapter 3, one of the ways the auditory system does this is by constructing a mental representation of presumed sound sources. The brain assembles the puzzle pieces of resolved partials into plausible images, streams, and textures that mirror the acoustic objects or groups of objects that exist in the real world.

As noted earlier, analyzing the auditory scene is a critical biological function, and one not limited to the processing of musical sounds. Parsing the acoustic scene is vital for tracking predators and prey, friends and foe. Any failure to correctly parse the acoustic scene reduces any animal's biological fitness. In the same way that deafness can reduce a person's chances of survival, a person who is confused by a complex acoustic scene is biologically endangered.

We know that dangerous situations or behaviors tend to evoke negative feelings. For example, there are excellent reasons why people should be fearful or apprehensive in a dark environment. Darkness renders an entire sensory system useless. Feelings of fear or trepidation motivate us to try to restore our vision by finding a light source or leaving the darkened space.

Similarly, we know that biologically advantageous behaviors tend to evoke positive feelings. For example, eating food is both life-sustaining and enjoyable. Since successful parsing of acoustic scenes is biologically advantageous, it would make sense that listeners would experience positive feelings when they successfully parse an acoustic scene. Conversely, failing to make sense of an acoustic scene might be expected to evoke negative feelings.

There is plenty of evidence that sensory systems evoke such positive and negative feelings. In chapter 3 we noted that visual obstruction leads to feelings of annoyance. Similarly, visual glare (such as reflections from the sun) is irritating, and watching an out-of-focus video can be quite unpleasant. In each of these cases, the capacity of the visual system to extract useful information from the environment is impeded. When the obstruction is removed, the glare eliminated, or the focus restored, we feel a palpable sense of relief or pleasure.

Similar experiences occur in audition. Hearing noise or static on the radio is sufficiently annoying to encourage us to retune or reposition the radio. When we leave a noisy environment, we often experience a feeling of relief. Switching from monophonic to stereophonic reproduction both improves the spatial resolution of the sound sources and simultaneously evokes pleasure. In the case of timbre and localization, momentary "drop out" has been experimentally demonstrated to reduce listener pleasure.[1] As we ascend higher up the auditory pathway, these low-level sensory pleasures and annoyances are joined by feelings linked to more complex perceptual processes, such as the mental representation of a perceptual scene. Once again, we seem to experience pleasure when we successfully assemble a coherent perceptual scene and experience irritation or annoyance when the scene fails to make sense. As the experimental research shows, the auditory system is not always successful in analyzing acoustic scenes. Sensory information can often be incomplete or ambiguous, and deciphering the world entails a certain amount of mental effort.

By auditory "success," we don't mean that the mental representation pieced together by perception accurately reflects the real world, since there is no independent way of verifying the accuracy. Rather, we mean that the resulting mental representation is coherent, fulfills expectations, and is consistent with information from other (equally fallible) senses. We can feel confident of success if there are no left-over puzzle pieces, if all the pieces fit snugly together, and if the overall picture makes sense. When this happens, the brain offers a reward that, if nothing else, reinforces the heuristic principles (and associated neural circuits) for successful auditory scene analysis.

When musicians follow the rules of voice leading, this helps to ensure that nominally independent things are perceived as truly independent. Conversely, when we apply the voice-leading rules in reverse, this helps to ensure that nominally fused things (like textural streams) are perceived as truly integrated. In short, the rules help to reduce perceptual ambiguity and so facilitate auditory scene analysis. If the rules of voice leading facilitate auditory scene analysis and if successful auditory scene analysis is rewarded by the brain, then following the rules of voice leading ought to contribute to the listener's experience of pleasure.

An apparent problem with the view I've proposed here is that composers seem to prefer music containing several concurrent parts. If we experience parsing success as pleasant, then why wouldn't musicians create solely monophonic music? In parsing the auditory scene, listeners would be more assured of success if the music consists of just a single line rather than many concurrent lines.

Experimental research, however, has established that listeners prefer multiple concurrent parts or instruments. In 2014, Brian Hurley, Peter Martens, and Petr Janata from the University of California, Davis, carried out two experiments that examined the effect of multiple parts on musical preferences. They found that as the musical scene density increases, listeners report greater enjoyment and are more likely to want the music to continue. They also examined the effect of staggered instrument entrances compared with static scene densities. They found that listeners enjoyed the staggered entrances much more than the static densities. The amount of enjoyment was greatest when a single instrument was joined by a second instrument. The amount of enjoyment was next greatest when two instruments were joined by a third instrument.[2] Since multiple sound sources are more difficult to parse than single sound sources, why are multiple sound sources more pleasurable for listeners?

A possible answer to this question is that the amount of pleasure evoked may be proportional to the perceptual difficulty posed by the scene. The harder the task, the greater the potential gratification. We typically experience more satisfaction from solving a difficult problem than solving a trivial problem.[3] Consider, by way of illustration, a parallel case in vision. A couple of decades ago a popular form of visual puzzle arose known as the random-dot stereogram, also called "magic eye" pictures.[4] These are three-dimensional images that are printed as a confusing array of random dots on a two-dimensional page. Seeing the 3D image can involve substantial mental effort, but if the viewer successfully resolves the picture, the effect is a delightful "rush."[5] This positive feeling is much greater than if one merely observed the actual three-dimensional object. Once again, the amount of pleasure seems linked to the difficulty of resolving random dots into a coherent visual image.

Similarly, it appears that multipart music raises significant challenges for listeners' auditory parsing capacities. A commensurately greater pleasure arises from successfully parsing several concurrent auditory streams compared with a single-stream monophonic texture. Especially when a scene is complicated, there is good reason to suppose that the emotional centers of the brain reward perceptual systems when the incoming information is successfully parsed.

Note that these puzzles are sensory rather than cognitive. We are not talking about the pleasures of solving riddles, jokes, or brainteasers. These are sensory enigmas rather than cognitive enigmas. For convenience, we might refer to them as *ear teasers*. An ear teaser is any complex musical texture or acoustic scene that nevertheless affords clear scene-analysis parsing, with a consequent pleasurable effect for listeners who successfully resolve the sensory challenge.

As a sensory process, notice that ear teasers can be sustained across time. Unlike a riddle that is posed and then (potentially) solved, in music, the acoustic scene represents a constant ongoing challenge. The visual equivalent would be a random-dot stereogram movie rather than a single static picture. The continuous nature of music means that the capacity of ear teasers for evoking pleasure is extended over time rather than being limited to a single "aha" moment.

Of course creating ear teasers requires careful construction. In the case of magic eye pictures, you can't just throw random dots on a page and expect viewers to find the result compelling. Similarly, you can't just throw notes on a page and expect listeners to enjoy the resulting acoustic scene. As we've seen, coherent auditory scenes exhibit certain structural regularities (including source location cues, harmonicity, Fitts's law, and onset synchrony, for example), so constructing such scenes requires careful planning and execution.

As musical notation matured in the late Middle Ages, European composers were able to arrange increasingly complicated acoustic scenes where multiple sound sources could be precisely coordinated in time. As the music became more complex, the challenge for listeners in parsing the acoustic scenes became increasingly daunting. What I am proposing is that early Renaissance polyphonists discovered an aesthetic effect akin to magic eye pictures: by challenging the listener's auditory parsing abilities, the potential for a pleasing effect could be magnified. However, this heightened pleasure would be possible only if listeners could successfully parse the more complicated acoustic scenes. Having increased the perceptual challenge, composers would need to take care in providing sufficient streaming cues. As in the construction of magic eye pictures, considerable skill is needed in constructing a coherent scene. In music, that skill was eventually codified in traditional part-writing rules. Following the perceptual principles is vital if listeners are to have any chance of successfully parsing complex acoustic scenes. An implication of this claim would be that at least some of the listeners who claim to dislike polyphonic music may fail to "get it" in the same way that some viewers fail to "get" magic eye pictures.

If it is true that music notation was an essential precursor—allowing the creation of complex ear teasers—this would imply that it would be difficult to achieve good parsable acoustic scenes if the music were exclusively improvised. In light of all the excellent improvised jazz in existence, one might think that this claim is absurd. However, most improvised jazz involves heterogeneous instrumentation, such as piano, saxophone, bass, and drums. There is little question that a listener will be able to parse the resulting acoustic scene since each musical line or textural stream exhibits a distinctive timbre. Timbral differentiation, the most underestimated principle in auditory scene analysis, provides one of the easiest ways to achieve independent auditory streams.

If you think talented improvisers can always produce parsable acoustic scenes, try improvising a work for three singers or four saxophones.

Once similar timbres are involved, principles like onset synchrony, pitch proximity, pitch co-modulation, and others loom large—whether the aim is to create independent auditory streams or produce a single coherent textural stream. For those who want to write a good arrangement for big band, knowledge of voice leading remains pivotal.

Of course, music notation is not the only means by which musicians can carefully arrange sounds. Over the past century, the organizational opportunities afforded by music notation have been supplemented by recording technology. Sound mixing and digital editing allow musicians to more precisely arrange or tailor specific acoustic scenes. In some respects, the digital audio studio is the new notation. However, even if traditional notational literacy became obsolete, the principles of auditory scene analysis would remain essential knowledge for good musicianship. Voice leading has not become obsolete in the age of studio-based music-making.

One acoustical property accessible to studio manipulation but normally outside notation is the manipulation of reverberation. In a reverberant hall, sounds ricochet back and forth between walls, ceilings, and other surfaces. The effect adds considerable complexity to the acoustic scene. In the case of speech, reverberation is the enemy of intelligibility. The vast majority of broadcast or recorded spoken voice is intentionally produced in acoustically dry environments in which reverberation is minimized or absent.

The contrast with music is striking. In music-making, we regularly seek out reverberation. Indeed, listeners find music recorded in dry conditions to be generally less pleasurable. Of course, with too much reverberation, we hear the result as sonic mush. So why do we like our music reverberant? And what is the right amount of reverb?

Ear teasing, I suggest, provides a plausible explanation.[6] Reverberation adds to the complexity of the acoustic scene. Like adding voices to a musical texture, it makes parsing more challenging. However, the added complexity is remarkably controlled. Reverberation does not add or subtract partials from the scene; it has minimal effect on harmonicity and on pitch co-modulation.[7] Its main effect is on sound onsets, where theoretically, individual echoes might be mistaken for new sounds. Tellingly, architectural acousticians design concert halls explicitly to avoid strong individual reflections (echoes). Musicians prefer (and acousticians therefore aim) to have a reverberant halo of multiple reflections without individual echoes being prominent. Reverberation might be expected to play havoc with sound localization, but in this case, biological evolution has been hard at work as evident in something called the Haas effect: in determining the location of a sound, the auditory system automatically suppresses reverberant sounds and relies exclusively on the initial (direct) sound.[8] In other words, although reverberation adds to the complexity of an auditory scene, moderate amounts of reverberation leave intact nearly all of the acoustical features affecting auditory scene analysis.

How much reverberation is too much? I would suggest that too much occurs when the reverb truly interferes with the parsing of the auditory scene. Notice that we tolerate much more reverberation in the case of a single solo voice or solo instrument than an ensemble.[9] Similarly, choral music in highly reverberant environments is least effective when the music is polyphonic, more effective when the music is homophonic, and most effective when the music is monophonic—consistent with an ear teaser theory of reverberation. In short, reverberation, like multipart music, adds complexity: complexity that might be expected to challenge the auditory system's ability to parse the scene, but complexity that is ultimately manageable.[10]

Finally, consider once again the phenomenon of stereo reproduction. In chapter 3 we discussed at length why stereo sounds better than mono reproduction. The effect isn't limited to sound reproduction; it is also evident in the hearing mechanism itself: deaf individuals outfitted with bilateral cochlear implants (one in each ear) are more likely to enjoy hearing music than deaf individuals wearing just one cochlear implant.[11] Interestingly, the pleasure of stereo listening is related to the number of sound sources present: the stereo effect is more pleasurable for multiple sound sources than for a single sound source. For example, the change from mono to stereo when listening to a solo clarinet isn't as dramatic as a similar change when listening to a brass quartet. Said another way, the pleasurable effect of stereo is greater as the density increases. The pleasure of listening to stereophonic sound bears the hallmarks of another enjoyable ear teaser.

Musical Scene Analysis

One area of musical scene analysis remains poorly understood, and that is our knowledge of what makes for good textural evolution—how musical scenes might change over time. We have empirical evidence for just one principle: listeners enjoy incremental textures where streams enter the musical scene one by one.[12] Film editing might provide a useful model for understanding how to design effective dynamic changes in musical scenes. Film editors and composers face similar challenges. Abrupt or dramatic changes of scene can be disorienting for both viewers and listeners. There is an art to using rapid changes of camera angles, positions, and perspective to add excitement without rendering the story-line confusing.[13]

Apart from the pioneering work of Ben Duane, we have almost no analyses of what kinds of musical textures have been used and how these textural practices have evolved over history. Like the history of film, it is probably the case that musical works have tended to exhibit increasingly complex auditory scenes over time—with more scene changes and more rapid transitions between scenes. However, these thoughts are purely speculative. We have almost no analyses to draw on. Both the analysis of musical texture and the analysis of the dynamics of changing musical scenes seem promising areas for future study.[14]

Plural Pleasures: Accommodating Multiple Goals

An underlying assumption in the discussion in this chapter has been that a motivating impetus in music-making is the pursuit of pleasure. Some readers may understandably be unnerved by this emphasis on the evoking of pleasure. Art has no predefined function, which means that it can be harnessed to serve any number of purposes, including no purpose at all. Sometimes art is successful because it educates us, inspires us, challenges us, disturbs us, or even insults us. But if art never offered any element of pleasure, it would cease to play much role in human affairs.[15] My assumption here is not that pleasure is the ultimate goal of music-making, or that the evoking of pleasure ought to be the ultimate goal for music-making. Instead, my assumption is that when certain sound combinations are pleasurable, they will inevitably be attractive to both listeners and musicians, and we shouldn't be surprised if much music-making succumbs to these allures.

When yielding to pleasure's attractions, it would be a mistake to assume that there is just one source of gratification. Ear teasing is not the only way musical sounds can be rendered enjoyable. I have chronicled many other sources of music-related pleasure in other publications.[16] Many of these hedonic goals can be pursued concurrently. However, sometimes two or more goals are inherently incompatible. For example, it is difficult for the same musical passage to evoke joyful energy in listeners while simultaneously evoking meditative relaxation. In chapter 8 we saw how competing goals might lead to compromises, such as synchronous onsets facilitating the intelligibility of choral lyrics while diminishing voice independence. As a further example, consider the characteristic choral practices found in much of sub-Saharan Africa.

Working at the Ohio State University, Aaron Carter-Enyi has documented some of the factors involved in the development of African harmony, especially in Nigerian choral music.[17] Most African languages are tonal languages; like Mandarin Chinese, words change meaning according to the pitch inflection used. For example, the Igbo language, spoken in southeastern Nigeria, has two main tones: high and low. A two-syllable utterance like *akwa* can be spoken with four different pitch combinations, each having a different meaning: HH (cry), LL (bed), LH (egg), and HL (cloth).

As is the case with all tonal languages, singing in the Igbo language can raise special difficulties if the melodic contour conflicts with the intended lyrics. Even when there is a conflict, much of the time the meaning of the words will be clear from the context.[18] However, problematic situations can still arise. When Western missionaries arrived in Africa, they were eager to teach converts the joys of four-part European hymn singing. As we've learned, one of the key principles in creating independent musical parts is to minimize semblant pitch movement by emphasizing oblique and contrary motion.

However, when Igbo singers sing in contrary motion, the vocalists may no longer be singing the same words. Carter-Enyi provides a particularly vivid example in the hymn "All Hail the Power of Jesus' Name." In Igbo, the translation is "Ọ̀hà kèlé íké Jésù" (All hail the power of Jesus). The pitch contour of the speech prosody (LLLHHHHL) nicely mirrors the arch-shaped melodic contour. However, in the bass voice, the LHLHHLHL pitch contour produces Ọ̀há kèlé ìkè Jésù (Trees hail the buttocks of Jesus). One immediately sees why parallel part-writing is popular throughout most of sub-Saharan Africa, especially among cultures employing tonal languages.[19]

Indeed, several African composers like Ekundayo Phillips have advised their compatriots to emphasize similar and parallel motion and avoid melismatic text setting.[20] Note, however, that not all African composers have been so eager to abandon the pleasures afforded by independent voice leading. Nigerian composer Laz Ekwueme has offered alternative advice: notably to consider the use of nonsense vocables in nonmelodic parts and to make use of staggered or canonic devices.[21] Notice that the goal of intelligible lyrics remains a priority for both Phillips and Ekwueme. However, Ekwueme recognizes that there may be opportunities to "have your cake and eat it too." If the alto, tenor, and bass "hum" or sing "ah," then the composer can have both intelligible lyrics (in the soprano) and simultaneously write perceptually independent voices, consistent with traditional voice leading. Similarly, musical rounds, canons, and other forms of imitative polyphony make it possible to maintain the compatibility of melody and lyrics while providing opportunities for independent part-writing.

The overarching point here is that organized sound affords many opportunities for evoking pleasure in listeners and that voice leading does not necessarily trump all other goals. Often the pursuit of multiple goals leads to inevitable compromises. In other cases, clever artists can figure out ways to reconcile seemingly contradictory goals.

Reprise

In this chapter, we have addressed the question of why listeners might experience clear part-writing as enjoyable. We know that listeners perceive adding streams to a texture as pleasurable. At the same time, we know that increasing the density of a texture makes it more difficult for listeners to successfully parse an acoustic scene. I have suggested that the brain rewards itself for successfully parsing auditory scenes and that the evoked pleasure is proportional to the scene's complexity.[22]

Inspired by the concept of the brain teaser, I have proposed the term *ear teaser* to refer to this phenomenon. An ear teaser is any complex musical texture or acoustic scene that nevertheless affords clear scene-analysis parsing, with a consequent pleasurable effect for listeners who successfully resolve the sensory challenge.

Voice leading offers a set of heuristics intended to help listeners successfully parse dense auditory scenes. For this reason, passages that are constructed according to voice-leading principles are commonly heard as sounding better than passages that ignore these principles. I have suggested that the preference for reverberant musical sounds (in contrast to speech) and stereo reproduction (in contrast to mono) similarly amounts to ear teasing phenomena.

Finally, I have emphasized that ear teasing is not the only game in town. There are many other ways by which musicians can create a pleasing sonic effect. Voice-leading practices may be modified, diluted, or jettisoned in order to permit the achievement of other goals deemed more important.

17 Conclusion

Music is an art whose basic building blocks are sounds and silences. Even the seemingly simple sound of a single flute tone turns out to be far from simple. Sound sources vibrate in several simultaneous modes, each of which produces a distinct partial. A passage for full orchestra generates an incredibly complex tapestry of frequencies, and the ability of listeners to make sense of this sonic cacophony is a stunning perceptual achievement.

Recognizing modes of vibration is not as useful to minds as recognizing vibrating objects. Consequently, one of the foremost tasks of the auditory system is to assemble partials into coherent auditory images that plausibly correspond to the actual (acoustical) sound sources. Since sources tend to generate sounds that extend over time, the auditory system also attempts to link partials temporally—forming sonic lines or streams.

In recent decades major advances have been made in our understanding of how the human auditory system goes about parsing the acoustical scene. The physical world of sound is reconstituted as a subjective auditory scene populated by perceptually conjectured sound images.

Twelve principles are known to be important in auditory scene analysis: *harmonic fusion* (where harmonically related partials are more likely to cohere), *toneness* (where the most tone-like sounds are harmonic complex tones in the region spanned by the bass and treble staves), *minimum masking* (where auditory interference is reduced when the sound energy is spread evenly across the basilar membrane), *continuity* (where contiguous sounds connect better than intermittent sounds), *pitch proximity* (where neighboring pitches are heard as better connected), *pitch co-modulation* (where parallel contours encourage fusion), *onset asynchrony* (where synchronous sounds tend to fuse), *limited density* (where three or fewer concurrent streams are easier to follow), *timbral differentiation* (where contrasting timbres contribute to perceptual independence), *source location* (where wide source separation aids perceptual independence), *attention* (arising from either willful thought or passive sonic capture), and *expectation* (where musical lines are easier to follow when listeners have prior knowledge allowing them to anticipate sound events).[1]

Auditory scene analysis is an evolved adaptation that serves an important biological function. We can't turn off the mind's disposition to parse auditory scenes. Listeners will attempt to parse all musical textures, even if the texture was created without regard to scene organization.

Pleasure

Like most essential biological systems, the auditory system offers rewards and punishments intended to encourage adaptive functioning.[2] I have proposed that auditory scene analysis affords two sources of pleasure. One source is a simple *parsing reward* for successfully resolving an auditory scene. (Conversely, when an auditory scene fails to be coherently parsed, the corresponding punishment is a feeling of confusion or irritation.) The second source of pleasure is linked to complexity. In speech, we normally follow just one conversation at a time, and people politely take turns speaking. In contrast to speech, the norm in music is multiple concurrent sounds. Strangely, although listeners experience increasing difficulty when tracking denser auditory scenes, people nevertheless prefer music constructed using multiple streams more than music constructed using just one or two streams.[3] This apparent paradox is reminiscent of recreational puzzle solving, where people take pleasure from deciphering challenging brain teasers. Accordingly, the second source of pleasure uses perceptual difficulty to amplify the first pleasure. I have referred to this "acoustical challenge" strategy as *ear teasing* and the resulting musical scenes as *ear teasers*. I suspect that ear teasers are more effective with music compared with speech because music liberates listeners from the need to extract dense linguistic meaning.

On the one hand, musicians raise the stakes of scene analysis by regularly creating highly complex acoustic scenes. On the other hand, musicians can provide helping hands that assist listeners in successfully parsing the more complicated scenes. Foremost among the helping hands are the conventional part-writing rules: failing to consistently follow these conventions can lead to streams that inexplicably drop in or drop out—situations that render the scene confusing. However, scene analysis research suggests other useful aids that have traditionally not been regarded as part of voice leading. Working at the McMaster Institute for Music and the Mind, Tanor Bonin, Laurel Trainor, Michel Belyk, and Paul Andrews have shown that consistent use of timbre and localization to clarify streaming is more pleasurable for listeners than when timbre and location cues drop in and out.[4] Another simple way to aid scene parsing is to introduce streams one at a time. Recall that Hurley, Martens, and Janata showed that introducing sound sources in succession is more enjoyable than simply presenting all of the sounds at once.

In identifying these sources of auditory pleasure, it is important once again to remind ourselves that these are not the only sources of musical enjoyment. Music can evoke pleasure in many other ways apart from the pleasures associated with auditory scene analysis.[5] Moreover, the pursuit of other goals (such as intelligible lyrics, improvisational freedom, range restrictions, rhythmic or harmonic patterns, or formal organization) may require some compromises that tend to reduce the coherence of some musical scenes.

Setting Scenes

In chapter 1, I defined voice leading as a codified practice that helps musicians craft simultaneous musical lines so that each line retains its perceptual independence for an enculturated listener. It should now be clear that a musical line is not necessarily the same as a musical instrument, voice, or part. To be sure, musical lines may be *auditory streams* that the mind regards as plausible acoustic sound sources. But musical lines can also be *textural streams* where several sound sources behave as a coordinated group.

It's not surprising that the traditional part-writing rules, honed in the heyday of Baroque polyphony, assume that the composer's goal is to render each part perceptually independent. But also implied in the rules is the reverse process by which musical instruments, voices, or parts can be integrated to form cohesive perceptual lines. If you want a set of disparate instruments to cohere as a single textural stream, the best results will arise when they employ synchronous onsets, move in parallel, play harmonically related pitches, are positioned close together in space, and include more than three instruments, for example. The underlying principles aren't simply tools for creating polyphonic music or Baroque-style chorales; they are tools that allow composers to construct and control any kind of musical texture, where the sounding resources are deployed and redeployed to make compelling dynamic musical scenes.

Whatever the musical style, composers effectively engage in *scene setting*, where a hierarchy of partials, auditory streams, and textural streams is assembled. Composers can control and manipulate the density, hierarchy, foreground/background relations, as well as the temporal evolution of these scenes. Traditional textures (e.g., tune-and-accompaniment, close harmony, polyphony, yodeling, homophony) all amount to different classes of auditory scenes.

When setting a musical scene, musicians are necessarily in partnership with listeners—since the acoustical world is ultimately experienced through its auditory twin. That is, musicians construct acoustic scenes, whereas listeners construct auditory scenes. Most parsing of the auditory scene is carried out automatically with the listener relying on a stream-of-consciousness mode of perception. However, astute listeners may be able to exercise some choice in either attending to the overall integrated texture (synthetic listening) or on the underlying sound components (analytic listening).[6] What we attend to depends on attentional effort, the complexity of the scene, individual listening skill, and (notably) experience with various kinds of sound environments and objects.

When a musical ensemble consists of a small number of heterogeneous instruments, acoustical scenes are nearly always easy to parse. For wholly improvised music, timbral differentiation and limited density compensate for a multitude of uncontrolled factors that might otherwise cause confusion. However, when similar timbres are used and three or more parts are present, much greater control is needed in order to craft acoustic scenes that listeners can successfully parse. That skill is embodied in the traditional canon of voice-leading rules.

A Revised Voice-Leading Canon

It bears reiterating that the purpose of this book is not to tell musicians how to create music. Instead, our goal has been to provide a sort of road map

Conclusion

that describes the consequences of choosing one path rather than another. Said another way, the aim has been to explain how the tools work, not to tell musicians what to create using those tools. Any musician will form his or her own aesthetic goals, and those goals may or may not include how listeners parse the resulting acoustic scenes.

For those musicians interested in creating independent musical parts or voices in a traditional tonal context, the extant auditory research offers an opportunity to revisit the traditional Baroque four-part chorale-style part-writing canon. It is possible to prune some redundant rules and to refine, clarify, and extend the part-writing advice. Traditionally, compositional canons are expressed in the form of hard-and-fast rules; the advice offered here is stated in the form of preference rules rather than decrees or commandments.[7]

If a musician chooses to create music in which two or more concurrent parts or voices are intended to be heard as perceptually distinct, then the following preferences should be observed:

1. *Toneness rule.* Prefer harmonic complex tones rather than inharmonic or aperiodic sounds.
2. *Compass rule.* Prefer the pitch region between about E2 and G5, centered near D4.
3. *Sustained sound rule.* Prefer continuous or sustained tones in close succession, with few long gaps or silences.
4. *Spacing rule.* Prefer wider spacing between the lower tones in a sonority.
5. *Tessitura rule.* Increasingly prefer wider spacing as the sonority becomes lower in overall pitch.
6. *Unisons rule.* Resist shared pitches between voices.[8]
7. *Octaves rule.* Resist the interval of an octave between two voices.
8. *Compound fifths rule.* Resist the interval of a perfect fifth (and its compound octave equivalents) between two voices.
9. *Harmonic fusion rule.* Resist unisons more than octaves, octaves more than perfect twelfths, perfect twelfths more than perfect fifths, and perfect fifths more than other intervals.
10. *Common tone rule.* If successive sonorities share a common pitch class, prefer to retain this as a single pitch within one voice.

11. *Conjunct motion rule.* If a voice cannot retain the same pitch from one sonority to the next, the preferred pitch motion is by step. Resist large melodic intervals.

12. *Leap lengthening rule.* When a large leap is unavoidable, long durations are preferred for either one or both of the tones forming the leap.

13. *Proximity rule.* Prefer writing parts that move to the nearest chordal tone in the next sonority.

14. *Crossing rule.* Resist the crossing of parts with respect to pitch.

15. *Overlap rule.* Resist "overlapped" parts in which a pitch in an ostensibly lower voice is higher than the subsequent pitch in an ostensibly higher voice.

16. *Leap away rule.* When large melodic intervals are used, prefer to assign it to the highest or lowest voice, and prefer to leap away from the other voices.

17. *Semblant motion rule.* Prefer nonsemblant over semblant motion between concurrent parts; that is, resist similar or parallel motions.

18. *Parallel motion rule.* If semblant motion is necessary, prefer similar motion over parallel motion.

19. *Oblique preparation rule.* When approaching unisons, octaves, fifteenths, twelfths, or fifths, it is preferable to retain the same pitch in one of the voices (i.e., approach by oblique motion).

20. *Conjunct preparation rule.* If it is not possible to approach unisons, octaves, fifteenths, twelfths, or fifths by retaining the same pitch (oblique motion), then step motion is the next preferred approach.

21. *Nonsemblant preparation rule.* Resist similar pitch motion in which the voices employ unisons, octaves, or perfect twelfths/fifths (e.g., when both parts ascend beginning an octave apart, and end a fifth apart).

22. *Perfect parallels rule.* Resist parallel unisons, octaves, twelfths, and fifths.

23. *Direct intervals rule.* When approaching unisons, octaves, twelfths, fifths, or fifteenths by similar motion, step motion is preferred preceding the upper pitch forming the interval.

24. *Asynchronous onsets rule.* Prefer asynchronous onsets for concurrent voices.

25. *Asynchronous preparation rule.* When approaching unisons, octaves, twelfths, or fifths, prefer asynchronous note onsets.

26. *Embellishments rule.* It is preferable to interpose embellishments (like passing tones) between successive sonorities.

27. *Embellishment preference rule.* Prefer embellishments using step motion (like passing tones, neighbor tones, suspensions and retardations) or no motion (pedal tones, repetitions and anticipations). Use embellishments involving leaps (like escape tones, appoggiaturas, and arpeggiations) less often.

28. *Doubled pitch-class rule.* Prefer adding embellishments to voices that double the pitch class of another concurrent voice.

29. *Embellishment refresh rule.* Prefer to add embellishments to voices that have not been embellished recently.

30. *Follow tendencies rule.* Prefer to resolve tendency tones in the expected (i.e., most common) manner.

Some common (though fallible) simplifications of the above *tendencies rule* include the following four more concrete situations:

31. *Chromatic resolution rule.* When a chromatically altered pitch is introduced, prefer the expected resolution.

32. *Chromatic backtracking rule.* When a chromatically altered pitch is introduced, don't backtrack by returning to the unaltered pitch.

33. *False relation rule.* Prefer to avoid two successive sonorities where a chromatically altered pitch appears in one voice but the unaltered pitch appears in the preceding or following sonority in another voice.

34. *Uncommon intervals rule.* Prefer to avoid an infrequently used melodic interval unless you plan to use lots of them (that is, unless you plan to take advantage of *dynamic-adaptive* expectations).

35. *Augmented intervals rule.* Prefer to avoid augmented melodic intervals unless you plan to use lots of them (i.e., take advantage of dynamic-adaptive expectations).

36. *Repetitive patterns rule.* Prefer to repeat distinctive melodic patterns, such as themes, motives, figures, subjects, or other sequences.

37. *Doubling rule.* Prefer to avoid doubling tendency tones, such as chromatic notes, the leading tone, or other tones that in context would evoke strong expectations.

Two points of caution are appropriate here. First, the revised canon I've offered is intended as advice for advanced students of music. These preference rules are not suitable for students learning part-writing for the first time (see the Afterword). Second, the study of voice leading remains a work in progress rather than a finished opus. With future research, the interpretations I have offered in this book will be corrected, augmented, or replaced.

Performance Considerations

Apart from arranging and composing, a broader understanding of voice-leading holds several practical repercussions for conductors and performers. For example, auditory scene analysis tells us that changing the tempo can change how listeners perceive musical lines. In general, the faster the tempo, the greater the number of perceived auditory streams. An accelerando doesn't just make the music sound more energetic or urgent; it makes the polyphony more complex. A faster tempo might change an orchestral arrangement from four textural streams to five textural streams, and conversely, an excessively slow tempo might reduce three textural streams to just two streams—potentially reducing the pleasure listeners experience from higher-density acoustic scenes.

Synchronizing onsets is the most important way that multiple instruments can amalgamate to form a textural stream. Apart from synchronizing onsets within one's textural group, there is value for one group to play slightly ahead or behind another textural group. Over the course of a work, individual ensemble performers should be alert regarding the different textural streams in which they participate. Depending on how instruments are assigned to different textural roles, conductors might even consider alternative seating arrangements that enhance scene parsing through localization cues. Barbershop quartets singers intuitively understand the value of localization cues when then they place their heads in close physical proximity.

For solo performers, increasing the durations of antecedent and consequent tones in a melodic leap (leap lengthening) will help ensure the perception of a single line. Leap lengthening is especially important when the accompanying musical scene is dense. Similarly, a soloist's rubato becomes more important as the accompaniment increases in density. That is, there are good perceptual reasons why a solo performer should either lag behind

Conclusion

the beat or push ahead of the beat when accompanied by a large ensemble. Even when playing in unison, a single musical part can be made to pop out by using vibrato, portamento, or glissandi (all forms of modulation or micromodulation).

Repetition or imitative melodic patterns make the music more predictable and so facilitate polyphonic listening. In short, taking repeats is likely to facilitate scene parsing especially when the music is complex.

Wider spacing in the bass decreases masking. Masking is also reduced when the overall sound energy is lower. So the close-position bass chords in Brahms's piano works will sound less muddy if played quietly.

Finally, consider one additional repercussion for performers mentioned in chapter 5. Recall that the high-voice superiority effect is caused by auditory masking, which arises from the mechanics of the basilar membrane. Throughout history and in nearly every documented culture, women's participation in music-making has been actively discouraged, restricted, or even forbidden. In the face of pervasive prejudice against women, who could have imagined that a low-level anatomical detail of human hearing probably had a decisive influence, allowing women to perform the most prominent of vocal roles.

Coda

The practical repercussions for composers and performers aside, perhaps the main value of scene analysis principles is in better understanding various aspects of musical organization. No research is definitive. Nevertheless, the auditory research reviewed in this book offers plausible explanations for a number of musical phenomena.

The research suggests how melodies are able to cohere as lines of sound, and identifies situations where a melody is likely to fracture or fall apart. The research explains why musical lines are dominated by the use of small intervals, and why large leaps tend to employ longer duration tones.

The research suggests why multipart music tends to be preferred over simple monophonic music. The research appears to explain why stereophonic reproduction is more enjoyable than monophonic reproduction. And it suggests a plausible explanation for why a moderate amount of reverberation might be enjoyed by music listeners, and how the optimum amount of reverb might be expected to change with the density of the musical texture.

The research suggests why wider harmonic intervals tend to occur between the lower voices, and why wide intervals are unnecessary when then chord has a high tessitura. The research explains why music-making favors pitches centered around middle C (262 Hz), rather than around the logarithmic center of hearing near 1,000 Hz. The research offers good reasons why musicians have tended to prefer pitched instruments and instruments that generate harmonic partials. There are good reasons why carillonneurs claim that bells are ill-suited to polyphonic music. The research suggests why musicians prefer instruments that are capable of sustained tones, while also explaining why damping mechanisms are used to stop sounds from continuing to vibrate. It explains why banjo and xylophone arrangements are faster paced than comparable arrangements for guitar or marimba.

The research explains the conditions when parallel fifths and octaves are preferred or shunned. It explains why musicians might avoid part-crossing and part-overlapping. The research suggests why embellishing tones might be useful and why some embellishing tones are more popular than others. The research accounts for seemingly oddball musical injunctions, like the false relations rule and the direct intervals rule. At the same time, it accounts for why some variants of the rules are more common than others. For example, the research explains why step motion to octaves is more important in the higher of the two voices forming the interval.

As we've seen, the research explains how simply changing tempo (without changing the notes) can result in notable changes in how a musical texture is perceived. Apart from the notated music, it also explains why performers tend to slow down when performing wide intervals—and why listeners prefer this. It also explains why leaps tend to leap away from the texture. As already noted, the research offers a cogent account for why melodies tend to be placed in the upper-most voice or part.

The research suggests why yodeling is compelling to listeners, and offers insights into how yodeling is physiologically possible. It explains why musical imitation might be favored in polyphonic music, and offers a plausible reason why some listeners have trouble "getting" polyphonic music.

The research provides a perceptually grounded basis for analyzing and understanding different types of musical textures, whether tonal, atonal, or acousmatic. It provides a useful point of departure for analyzing patterns of orchestration. The research provides a unifying framework in which all kinds of musical styles and genres can be described and compared, and the research offers an aesthetic rationale for at least one potent form of music-induced pleasure.

The research indicates that individual listening experience and cultural background play a formative role when listening to complex textures. This same research suggests why music from an unfamiliar culture can sound aimless or meandering. The research demonstrates a notable malleability or plasticity in the auditory system. For example, the research shows that pitch perception itself is shaped by the instrument a musician plays. The objects we apprehend (like chords or car horns) depend to a considerable extent on environmental exposure.

The research suggests that it is not just top-down expectations that are learned from exposure to some sound environment. Some bottom-up elements of the auditory system also appear to be exquisitely sensitive to the sonic environments in which we are immersed. While the disposition for human minds to parse auditory scenes is surely an evolved innate behavior, the means by which this is achieved likely involves a mix of innate and learned mechanisms.

To the extent that auditory scene analysis is linked to learned patterns, stylistic norms likely influence how listeners parse a given musical scene. This implies that the perception of music is at least partially embedded in a sort of feedback loop in which musical culture itself helps shape how we perceive music. This suggests that the creative future of music-making may be more open than closed.

Afterword

Educators make a helpful distinction between two ways of knowing something: *procedural knowledge* (knowing how) contributes directly to the exercise of a skill; *declarative knowledge* (knowing that) entails conscious understanding, such as knowing what makes a skill effective. Over the years, I have taught voice leading using both procedural and declarative approaches. That is, I have taught voice leading using both the traditional rules-oriented approach and a principles-oriented approach. There are good reasons to expect that a principles-oriented approach might be preferred over an approach that stresses rote memorization and mechanical application. By understanding the underlying principles, one might well expect novice students to produce better part-writing. However, this expectation has not been borne out in practical classroom experience.[1] The perceptual principles described in this book are typically too abstract and complicated for music students who are just beginning the craft of part-writing.

In this book, I have presented the core voice-leading canon in the form of preference rules rather than using the more traditional fixed commandments approach. I believe that presenting the canon as a set of recommendations better reflects the many competing concerns that musicians must often reconcile when composing or arranging. Once again, however, practical classroom experience suggests that preferences rules are ill suited to good pedagogy. When *avoid parallel fifths* is offered as advice rather than as a dictum, the tendency is for students to be cavalier about occurrences of parallel fifths. If parallel fifths are regarded as simply "wrong," then students are much more motivated to backtrack and revise what they have already written.

In complicated ventures like voice leading, beginning students need strong guidelines. Black and white do's and don'ts help to avoid confusion and build confidence. Having explored a number of teaching approaches, I have developed a healthy respect for the rules of voice leading as traditionally taught. The rules are succinct, practical, relatively easy to learn, and nearly always lead to musically acceptable results.

I have found that once a student has mastered the practice of voice leading, a subsequent account of the psychological origin of the rules becomes more meaningful and more satisfying. In short, despite the attractions of introducing voice leading from the principles described in this book, I urge teachers to resist this temptation.

Of course, asking students to follow a set of rules without offering any cogent reason has the predictable consequence of generating suspicion and resistance. Throughout history, musicians have occasionally rebelled against an educational approach that seems wholly authoritarian. There are musically appropriate situations where the rules can be applied in reverse (such as when forming textural streams). Savvy music students will be well aware that many musical passages appear to violate the rules, and consequently they will rightly question why they should follow such an old-fashioned system. I have found that a useful place to begin is to tell new students that the rules that they are about to learn have an underlying logic. The rules may seem arbitrary, but they are not. After they have gained some mastery in applying the conventional rules of voice leading, we can then discuss where the rules might come from, why they make sense, why listeners might find the results enjoyable, and why learning a seemingly arcane Baroque practice holds value for any style of music-making. It is that story which I have tried to summarize in this book.

Acknowledgments

My interest in the perceptual foundations of voice leading began on July 15, 1981, when I heard a presentation given by McGill University psychologist Al Bregman. His presentation was at a small conference on music and psychology organized by Lola Cuddy at Queen's University in Kingston, Canada. Bregman's work on auditory scene analysis was my introduction to how psychology might account for many musical practices. The effect was electrifying, and that spark has inspired me for the past thirty-five years. For his groundbreaking work, I am deeply grateful to Professor Bregman.

Much of the research I describe in this book was done by hearing scientists and auditory psychologists who had little interest in music. But a handful of musician-researchers have focused explicitly on musical issues. For their important work and friendship, I am indebted to Jay Dowling, Leon van Noorden, Stephen McAdams, James Wright, Rudolph Rasch, Bret Aarden, Paul von Hippel, and Mark Yeary.

Above all, I thank the many colleagues and friends who provided stimulating conversation, correspondence, critiques, and encouragement. Foremost, I offer my thanks to Peter Schubert and David Temperley. In addition, my thanks go to Claire Arthur, David Butler, Peter Cariani, Nat Condit-Schultz, Zohar Eitan, Robert Gjerdingen, Henkjan Honing, Elizabeth Hellmuth Margulis, Elizabeth Marvin, Eugene Narmour, Joy Ollen, Richard Parncutt, Nancy Rogers, Daphne Tan, and the late David Wessel. I am especially grateful to those who took the time to read multiple drafts of this book. To all of these individuals, my heartfelt thanks.

Notes

1 Introduction

1. No list of citations here could do justice to the volume and breadth of writings on this subject. The following list is intended to be illustrative by including both major and minor writers, composers and noncomposers, didactic and descriptive authors, different periods and nationalities: Aldwell and Schachter (1989); Berardi (1681); Fétis (1840); Fux (1725); Hindemith (1944); Horwood (1948); Keys (1961); Laitz (2012); Morris (1946); Parncutt (1989); Piston (1962/1978); Rameau (1722); Riemann (1903); Roig-Francoli (2010); Schenker (1906); Schoenberg (1911/1978); Stainer (1878); and many others.

This book addresses voice leading from the perspective of perceptual and cognitive psychology. There exist other approaches to the study of voice leading, notably the mathematical and systematized voice leading associated with neo-Riemannian theory. This includes work by Richard Cohn, Henry Klumpenhouwer, and Dmitri Tymoczko (among others), often inspired by the pioneering work of David Lewin (see, e.g., Callender, 1998; Cohn, 1996, 1998, 2012; Lewin, 1987; Klumpenhouwer, 1994; Tymoczko, 2008, 2011). An important connection to the current work is the concept of *voice-leading parsimony* (Cohn, 1996; Tymoczko, 2008); see chapter 6.

2. See Bregman (1990), McAdams and Bregman (1979), van Noorden (1975), Wright (1986), and Wright and Bregman (1987).

2 The Canon

1. Von Dadelsen (1988).

2. Pedagogical works that focus on sixteenth-century modal counterpoint include Gauldin (1985), Jeppesen (1939/1963), and Schubert (1999, 2007). Works that emphasize eighteenth-century Bach-style counterpoint include Benjamin (1986), Parks (1984), Schubert and Neidhöfer (2005), and Trythall (1993). Examples of works that emphasize a harmonic approach include Piston (1962/1978). Aldwell and Schachter (1989) adopt what might be called a voice-leading approach to harmony.

3 Sources and Images

1. MacDougall and Moore (2005).

2. *Ossicles* rhymes with "popsicles"; *stapes* is pronounced *STAY-peez*; *cochlea* is pronounced either *COCK-lee-ah* or *COKE-lee-ah*.

3. Georg von Békésy (1943/1949, 1960), pronounced *GEH-org fawn BEH-keh-she*.

4. The term was coined by Albert Bregman (1990).

5. See Bregman (1978, 1990), Bregman and Campbell (1971), Bregman and Pinker (1978), and McAdams and Bregman (1979). See also Carlyon (1991, 1992, 1994), Patterson (2000), and Patterson, Robinson, Holdsworth, McKeown, Zhang, and Allerhand (1992).

6. If the mono signal is produced by two loudspeakers, then partials W, X, Y, and Z will be present in both speakers equally.

7. The notion of the theater of the mind was first described by the seventeenth-century French philosopher René Descartes.

8. Bregman and Campbell (1971).

9. Once again, unresolved partials also play a role in forming auditory images, but only as a group.

4 Principles of Image Formation

1. Duifhuis, Willems, and Sluyter (1982); Scheffers (1983).

2. Brunstrom and Roberts (1998).

3. Traditionally this phenomenon has been called *tonal fusion* rather than *harmonic fusion*; see DeWitt and Crowder (1987). The term *harmonic fusion* is preferred here since it isolates a single cause of fusion—namely, coinciding harmonics. By contrast, the term *tonal fusion* is more ambiguous, suggesting that other reasons might cause two tones to cohere together, such as synchronous onsets or coordinated micromodulation.

4. Stumpf (1890); DeWitt and Crowder (1987).

5. For a psychoacoustic definition of tonality, see ANSI (1973).

6. Attneave and Olson (1971); Ohgushi and Hato (1989); Semal and Demany (1990).

7. Terhardt, Stoll, and Seewann (1982a, 1982b).

8. The Terhardt, Stoll, and Seewann model calculates a *pitch weight* for both pure and complex tones, and these pitch weights may be regarded as an index of the pitch's clarity, and therefore a measure of toneness. This model makes a distinction

between pitches evoked by pure tones (so-called spectral pitches) and pitches evoked by complex tones (so-called virtual pitches). For spectral pitches, sensitivity is most acute in the *spectral dominance region*, a broad region centered near 700 Hz. However, for virtual pitches, the greatest pitch weights arise when the evoked or actual fundamental lies in a broad region centered near 300 Hz—roughly D4 immediately above middle C.

9. From Huron (2001), using the method of Terhardt, Stoll, Schermbach, and Parncutt (1986). Toneness values are actually maximum "pitch weight" values in Terhardt's model. Pitch weights were calculated for each tone where the most intense partial was set at 60 dB. The solid curve shows the calculated maximum pitch weight of the most prominent pitch for sawtooth tones. The dotted curves show changes of calculated pitch weight for the recorded orchestral tones. For tones below C6, virtual pitch weight is greater than spectral pitch weight; above C6, spectral pitch predominates—hence the abrupt change in slope. The figure shows that changes of spectral content have little effect on the region of maximum pitch weight. Figure is from Huron (2001). Recorded tones from Opolko and Wapnick (1989). Spectral analyses kindly provided by Gregory Sandell (1991a).

10. Huron and Parncutt (1992).

11. Other metaphors are used in other cultures, notably the pitch-size metaphor.

12. Before leaving the subject of pitch, I should add that there is at least one additional attraction of pitch for musicians: pitch provides an opportunity to create harmony. However, harmony is not the subject of this book.

13. Once again, unresolved partials also play a role in forming auditory images, but only as a group.

5   Auditory Masking

1. The exception is so-called *destructive interference*. This occurs when two sounds of identical frequency are out of phase with each other. One sound cancels the energy of the other.

2. Békésy (1943/1949, 1960).

3. The basilar membrane does not span the full length of the cochlea, which is the reason that its length is shorter than the 4 centimeters noted in chapter 3.

4. Fletcher (1940, 1953).

5. Greenwood (1961b, 1990).

6. The width of critical bands is roughly intermediate between a linear frequency scale and a logarithmic frequency scale for tones below about 400 Hz. For higher frequencies, the width becomes almost completely logarithmic. When measured in

hertz, critical bandwidths increase as frequency is increased; when measured in semitones (log frequency), critical bandwidths decrease as frequency is increased.

7. Piston (1962/1978).

8. Noll (1993); Brandenberg (1999).

9. Marie and Trainor (2013); Trainor, Marie, Bruce, and Bidelman (2014); see also Chon and Huron (2014).

10. Another example is *faux bourdon*.

11. In their original publication, Plomp and Levelt (1965) estimated that pure tones produce maximum sensory dissonance when they are separated by about 25 percent of a critical bandwidth. However, their estimate was based on a critical bandwidth that is now known to be excessively large, especially below about 500 Hz. Greenwood (1991) has estimated that maximum dissonance arises when pure tones are separated by about 30 to 40 percent of a critical bandwidth.

12. See Greenwood (1961a, 1990, 1991); Kameoka and Kuriyagawa (1969a, 1969b). A further replication can be found in Iyer, Aarden, Hoglund, and Huron (1999).

13. See Van de Geer, Levelt, and Plomp (1962).

14. Plomp and Levelt (1965) originally referred to the phenomenon as "tonal consonance." Kameoka and Kuriyagawa (1969a, 1969b) pointed out that the primary phenomenon is one of dissonance rather than consonance (its presumed reverse). What I have been calling "sensory irritation" is commonly referred to in the published research as "sensory dissonance."

15. I have used Kaestner's (1909) data principally because he reports independent results for the unison and octave intervals. This will prove useful when we examine the role of tonal fusion in Bach's use of harmonic intervals. See below.

16. Data here relate to the upper two voices from Bach's three-part Sinfonias (BWVs 787–801). These voices were selected in order to minimize intervals larger than an octave for which consonance data were not available. See Huron (1991b) for a detailed explanation.

17. Bregman (1990).

18. Huron (1991b; see also Huron, 1994).

19. See Huron (1991b).

20. I examined several relationships in Huron (1991b). When perfect intervals are excluded from consideration, the correlation between Bach's interval preference and the sensory dissonance Z-scores (derived from Kaestner's data) is –0.85. Conversely, if we exclude all intervals apart from the perfect intervals, the correlation between Bach's interval preference and the tonal fusion data is –0.82.

6 Connecting the Dots

1. See Houtgast (1971, 1972, 1973); Thurlow (1957); Warren, Obusek, and Ackroff (1972).

2. Neisser (1967).

3. A number of experiments have attempted to estimate the duration of echoic memory. These measures range from less than 1 second (Treisman & Howarth, 1959) to less than 5 seconds (Glucksberg & Cowen, 1970). But typical measures lie near 1 second in duration (Crowder, 1969; Guttman & Julesz, 1963; Rostron, 1974; Treisman, 1964; Treisman & Rostron, 1972). Kubovy and Howard (1976) concluded that the lower bound for the half-life of echoic memory is about 1 second.

4. Using very short tones (40 msec duration), van Noorden (1975) found that the sense of temporal continuation of events degrades gradually as the inter-onset interval between tones increases beyond 800 msec. Further evidence supporting the importance of tone duration in streaming is found in Beauvois (1998). See also Bregman, Ahad, Crum, and O'Reilly (2000).

5. Huron (2001).

6. Johnson (2010).

7. Miller and Heise (1950); Heise and Miller (1951).

8. Including Bozzi and Vicario (1960), Vicario (1960), Schouten (1962), Norman (1967), Dowling (1967), van Noorden (1971a, 1971b), and Bregman and Campbell (1971). Several of these researchers worked independently, without knowledge of previously existing work. Of these pioneering efforts, the most significant works are those of Dowling (1967) and van Noorden (1975).

9. Van Noorden's terms differ from those used here: he called the *trill boundary* the *fission boundary* and the *yodel boundary* the *temporal coherence boundary*.

10. Dowling (1967). The term *stream* here is anachronistic: Bregman and Campbell didn't coin the term until 1971.

11. Dowling (1967). Dowling's work predated van Noorden, so the boundaries used by Dowling were those of Miller and Heise (1950).

12. Several authors have observed the pervasive use of small intervals in the construction of melodies, including Ortmann (1926), Merriam, Whinery, and Fred (1956), and Dowling (1967).

13. Deutsch (1975). Similar experiments illustrating the problem of part-crossing were carried out by van Noorden (1975). Further experiments were carried out by Butler (1979). Another demonstration of the problem of part-crossing is evident in

experiments by Jay Dowling (1973). In chapter 10 we will discuss an experiment by Dowling where he interleaved two melodies.

14. In the Gestalt school of psychology, the crossed trajectories would be considered the simpler figure and would be the preferred perception. The "bounced" perception in audition contradicts the Gestalt principle of "good continuation." In sequences of pitches, listeners tend not to extrapolate future pitches according to contour trajectories. Rather they interpolate pitches between pitches deemed to be in the same stream (Steiger & Bregman, 1981; Tougas & Bregman, 1985; Ciocca & Bregman, 1987, summarized by Bregman, 1990). Similar effects have been observed in speech perception (Darwin & Gardner, 1986; Pattison, Gardner & Darwin, 1986).

15. Huron (1991a).

16. J. S. Bach's keyboard fugues are peppered with incoherent voice reassignments where the range for a given voice changes over the course of the piece. For example, in the Fugue in F major BWV 540, at one point the alto voice effectively disappears; the tenor shifts to become the new alto, and a new tenor part appears between the former tenor and the bass (measure 110). Similarly, in the exposition of the five-part fugue in F minor BWV 534, the final statement of the subject appears in the uppermost part; however, after just three notes, this nominally "soprano" voice shifts down to the alto (measure 17). In many cases, these range reassignments can be clearly attributed to the aim of minimizing part-crossing.

17. Fitts (1954).

18. Korte (1915).

19. The similarity between Miller and Heise's trill results and Korte's third law of apparent motion in vision was independently noted by van Noorden (1975) and Shepard (1981).

20. The idea that both Korte's third law of apparent motion and the yodel boundary in auditory streaming can be attributed to Fitts' law was first proposed by Huron and Mondor (1994). See also Huron (2001).

21. For additional perspectives on the concept of musical motion see Gjerdingen (1994), Todd (1995), and Todd, Cousins, and Lee (2007).

22. This result was described in Huron (2001).

23. See Temperley (2007, 2014). Temperley found that the correlation between interval size and the duration of the first note forming the interval is +0.88, while the correlation between interval size and the duration of the following note is +0.51. For other perceptual repercussions of melodic leaps, see von Hippel (2000) and von Hippel and Huron (2000).

24. Sundberg, Askenfelt, and Frydén (1983).

25. For further information regarding the physiology of the voice, see Sundberg (1987).

26. Helmholtz (1863).

27. See Bregman and Doehring (1984); McAdams (1982, 1984, 1989).

28. McAdams (1982, 1984, 1989).

29. McAdams (1989).

30. Huron (1987, 1989a).

7   Preference Rules

1. Lerdahl and Jackendoff (1985); Temperley (2004).

2. See, for example, Price (1984).

3. Schutz, Keeton, Huron, and Loewer (2008); see also Johnson (2010).

4. Plomp and Levelt (1965); Huron and Sellmer (1992).

5. Huron (1991b).

6. Retaining common tones also has the effect of increasing oblique contrapuntal motion. It may be that the principal benefit of the common tone rule is that it reduces similar or parallel motions (i.e., reduces pitch co-modulation).

7. Temperley (2007, 2014); see also work by Huron and Mondor reported in Huron (2006b).

8. Bregman (1990).

9. Huron (1991a).

10. Huron (1989a).

8   Types of Part-Writing

1. Bregman and Pinker (1978). Onset synchronization might be regarded as a special case of amplitude co-modulation. Apart from the crude coordination of tone onsets, the importance of correlated changes of amplitude has been empirically demonstrated for much more subtle amplitude deviations. For example, Bregman, Abramson, Doehring, and Darwin (1985) have demonstrated that the coevolution of amplitude envelopes contributes to the perception of tonal fusion.

2. Hirsh (1959).

3. Saldanha and Corso (1964); Strong and Clark (1967).

4. Beranek (1954/1986).

5. Rasch (1978, 1979, 1988).

6. Rasch (1988).

7. Rasch (1981).

8. Huron (1993a).

9. Huron (1989c).

10. Tatsuoka and Tiedeman (1954).

11. Vos (1995).

12. Huron (2008).

13. Huron (2008).

14. From Huron (1989b).

15. Parncutt (1993). Parncutt's chords were constructed using tones with octave-spaced partials (Shepard tones).

16. Schoeffler, Stöter, Bayerlein, Edler, and Herre (2013). Also, see Huron and Fantini (1989) for an example of the musical implications of limited density.

17. Estes and Combes (1966); Schaeffer, Eggleston, and Scott (1974); Gelman and Tucker (1975).

18. Strauss and Curtis (1981).

19. In English: "one, two, three, many" phenomenon. Descoeudres (1921).

20. Hindemith (1944); see also Mursell (1937).

21. Huron (1989a).

22. Huron (1989a).

23. See, for example, the discussion in Kostka, Payne, and Almén (2012, chap. 13).

24. Another suggestion might be offered. Homophonic textures often contain five or more parts. When a composer employs such large numbers of concurrent parts, there is a tendency for the individual component tones to be lost, while the listener may be encouraged to hear "synthetic" perceptions such as chords. This account raises the question of why a composer would follow the rules of voice leading when constructing such thick textures. See chapters 13 and 14.

25. The role of timbre on auditory streams was demonstrated by van Noorden (1975) and Wessel (1979) and tested experimentally by McAdams and Bregman (1979); Bregman, Liao, and Levitan (1990); Hartmann and Johnson (1991); and Gregory (1994).

26. Wessel (1979).

27. Iverson (1992).

28. Erickson (1975) has identified a number of ways in which timbre is used to distinguish musical elements. Rössler (1952) carried out an extensive analysis of the influence of organ registration on the perceptual independence of polyphonic and pseudo-polyphonic lines. Rössler's results underscore the role of timbre in the fission and fusion of auditory streams.

29. Chon and McAdams (2012).

30. Sandell (1991b).

31. Divenyi and Oliver (1989); Huron (1991c).

32. Haas (1951).

33. Martin (1965); Root (1980).

9 Embellishing Tones

1. Hindemith (1944) called them *nonchordal tones*; Piston (1962/1978) referred to them as *nonharmonic tones*; Aldwell and Schachter (1989) call them *figuration tones*; and Horwood (1948) and others have dubbed them *unessential tones*.

2. Meyer (1956).

3. Narmour (1990, 1992).

4. Huron (2006b).

5. Other accounts of embellishing tones highlight the relationship between stable and unstable tones. Embellishing tones typically introduce an element of psychological instability that is then resolved by movement to some more stable tone. A plausible account may be found in the so-called cognitive anchoring of unstable tones to stable ones, as demonstrated by Krumhansl (1979) and investigated more thoroughly by Bharucha (1984).

Yet another account of embellishing tones has been proposed by James Wright and Albert Bregman (1987). Wright and Bregman suggested that the dissonances created by nonchordal tones are dampened perceptually when the tones forming the dissonance are strongly segregated into independent auditory streams. When two concurrent tones are captured by independent streams, their potential dissonance is suppressed or neutralized. Thus, the degree to which a major seventh harmonic interval is perceived as dissonant is thought to depend on how well the constituent tones are integrated into their respective horizontal voices. When the segregation between possible streams is weak, any vertical dissonances are perceived as especially dissonant. Wright (1986) has proposed that the potential dissonance arising from nonchordal notes is minimized by ensuring good streaming. He further suggested that the observed increase in dissonance over the course of the history of

Western music is reflected in the manner by which dissonant intervals are prepared: over time, dissonances have been heightened by the dual practices of increasing the onset synchronization of the tones forming the dissonant interval and by decreasing the use of antecedent and consequent step motion. In short, much of the historical increase in musical dissonance is attributable to the weakening of auditory streaming. For a review and critique of this theory see Huron, 1991c.

6. Wright (1986).

7. Huron (2007).

8. Huron (2007).

10 The Feeling of Leading

1. Bregman (1990).

2. Bregman had little to say about "top-down" principles in auditory scene analysis except that expectation plays an important role.

3. For a review of this research, see Huron (2006b).

4. Aarden (2002, 2003).

5. Dowling (1973).

6. The concept of analytic listening is discussed in chapters 13 and 14. As used by hearing scientists, analytic listening is quite narrowly defined. It should not be confused here with the general ability of musicians to listen to a piece of music from an "analytic" perspective—such as carrying out a harmonic analysis on the fly or following the signposts of a sonata-allegro form.

7. Dowling (1973).

8. See Aarden (2002, 2003); Carlsen (1981); Huron (2006b); and Narmour (1991, 1992). See Huron and Margulis (2010) for a review.

9. For an extended discussion, see Huron (2006b, chap. 12).

10. See Huron (2006b) for further discussion.

11. For a more detailed discussion, see Huron (2006b).

12. See Saffran, Johnson, Aslin, and Newport (1999).

13. See Saffran et al. (1999); Desain, Honing, and Sadakata (2003); Aarden (2002, 2003).

14. The probabilities listed in table 10.1 were calculated from several thousand German folksongs and represent more than a quarter of a million note transitions.

15. See Huron (2006b).

16. See Huron (2006b).

17. Incidentally, unlike schematic and veridical expectations, dynamic-adaptive expectations are available to listeners when exposed to any musical work, from the most familiar work to a novel musical work from an unfamiliar culture.

18. Gjerdingen (2007).

11 Chordal-Tone Doubling

1. See Aarden (2001).

2. Aarden and von Hippel (2004); see also Huron (1993b).

3. When the lights are arranged horizontally, color-blind drivers follow the rule: *Stop when the leftmost light is on.* Unfortunately, while most horizontal traffic lights conform to this arrangement, not all do, much to the despair of color-blind drivers.

4. Aarden and von Hippel (2004).

12 Direct Intervals Revisited

1. Laitz (2012, p. 52).

2. Benjamin, Horvit, and Nelson (2007, p. 235).

3. Kostka, Payne, and Almén (2012, p. 77).

4. Roig-Francoli (2010, p. 89).

5. Clendinning and Marvin (2010, p. 262).

6. Aldwell, Schachter, and Cadwallader (2010, p. 109).

7. Arthur and Huron (2016).

8. Huron (1989b).

9. This result arises from a reanalysis of the data collected by Arthur and Huron, but was not reported in their original publication.

13 Hierarchical Streams

1. In the Canada of my childhood, the chord produced by a train horn was a root-position minor triad.

2. Gjerdingen (1994).

3. Cambouropoulos (2008).

4. Hall and Pastore (1992).

5. Cusack, Deeks, Aikman, and Carlyon (2004).

6. Yeary (2011).

7. Similar hierarchical texture trees have been produced by Duane (2008).

8. A musical stream might be distinguished, for example, as independent of other concurrent sounds such as the sound of audience rustling. Duane also points to the famous example of two bands "colliding" in the second movement of Charles Ives's *Three Places in New England*. Here the texture may be best regarded as containing two musical streams rather than one.

9. Duane (2013).

10. Sacks (2007, p. 113).

11. Sacks (2007, p. 114).

12. Sacks (2007, p. 113).

14 Scene Setting

1. McNabb (1983).

2. For an example applied to "spectral" music, refer to the analysis of Ligeti's *Continuum* in Cambouropoulos and Tsougras (2009).

3. Landy (2007), p. 30; Bodin (1997), p. 223.

4. Schubert (1999), pp. 23–24.

5. My own analysis.

6. For a similar critique, see Thompson and Schellenberg (2006), p. 102.

15 The Cultural Connection

1. One known culture did not have fire-making skills: the indigenous peoples of Tasmania. However, archeologists have established that Tasmanians at one time did have fire, but subsequently they lost the skill.

2. Preisler (1993); Seither-Preisler et al. (2007); Schulte et al. (2002).

3. Seither-Preisler et al. (2007).

4. Schulte, Knief, Seither-Preisler, and Pantev (2002).

5. Schneider, Sluming, Roberts, Bleeck, and Rupp (2005).

6. Chorost (2005).

7. Remez, Rubin, Pisoni, and Carrell (1981).

8. Rubin (1915).

9. Pressnitzer and Hupé (2005).

10. Woodrow (1909).

11. Patel, Iversen, and Ohgushi (2004); Iversen, Patel, and Ohgushi (2008).

12. Patel et al. (2004); Iversen et al. (2008).

13. Incidentally, the most successful computer programs for auditory scene analysis make use of machine learning rather than programmed principles. See, for example, Smaragdis, Raj, and Shashanka (2007); Smaragdis, Shashanka, and Raj (2009).

16 Ear Teasers

1. Bonin et al. (under review).

2. Hurley, Martens, and Janata (2014).

3. Danesi (2002).

4. The first random-dot autostereogram was created in 1939 by Boris Kompaneysky.

5. Mikulas (1996).

6. In making this argument, I don't mean to suggest that ear teasing is the exclusive appeal of reverberation. For example, reverberation also typically reduces roughness and adds "extensity" or "volume" (S. S. Stevens).

7. Partials can sometimes disappear when reverberation causes acoustic nodes. The smearing of time can effectively increase oblique contrapuntal motion for short periods.

8. Haas (1951/1972).

9. Jazz flutist Paul Horn famously recorded an album of solo flute in the Taj Mahal with its extraordinary reverberation time of 28 seconds. It would be difficult for a performance of a three-part fugue to be effective under such conditions.

10. Once again, in making this proposal, I don't wish to imply that this is the only reason why listeners enjoy reverberant sounds.

11. Veeksman, Ressel, Mueller, Vischer, and Brockmeier (2009).

12. Hurley, Martens and Janata (2014); also Huron (1990, 1992).

13. Dmytryk (1984).

14. See Duane (2013) for preliminary efforts in this direction.

15. Huron (2006b, 2013).

16. For a list of possible sources of music-related pleasure, refer to Huron (2015b).

17. Carter-Enyi (2014).

18. Schellenberg (2012).

19. Carter-Enyi, Aina, Carter-Enyi, Huron, and Gomez (2013).

20. Phillips (1953).

21. Ekwueme (1974).

22. For an alternative view (one that emphasizes the ambiguity of auditory scenes), see Pressnitzer, Suied, and Shamma (2011).

17 Conclusion

1. This list of principles is not exhaustive. Other factors (not discussed in this book) include *offset synchrony* (the tendency for sounds with coincident termination moments to stream together), and *amplitude co-modulation* (where both positively and negatively correlated changes in amplitude contribute to stream fusion).

2. See Huron (2001, pp. 56–58). See also Bonin, Trainor, Belyk, and Andrews (under review).

3. Hurley, Martens, and Janata (2014).

4. Bonin et al. (under review).

5. Descriptions of other sources of pleasure can be found in Huron (2005, 2006b, 2013, 2015a, 2015b); Schellenberg, Corrigall, Ladinig, and Huron (2012).

6. Cusack, Deeks, Aikman, and Carlyon (2004).

7. Following Lerdahl and Jackendoff (1985); Temperley (2004).

8. Traditionally, an exception may be made for chords that end a phrase or musical work. This exception does not derive from the principles of auditory scene analysis. However, notice that this exception is consistent with the idea that an appropriate ending gesture might have the scene collapse to a fewer number of auditory streams, or even just one.

Afterword

1. See Huron (2006a) for further discussion.

References

Aarden, B. (2001). *An empirical study of chord-tone doubling in common era music* (Unpublished master's thesis). Ohio State University, Columbus.

Aarden, B. (2002). Expectancy vs. retrospective perception: Reconsidering the effects of schema and continuation judgments on measures of melodic expectancy. In C. Stevens, D. Burnham, G. McPherson, E. Schubert, & J. Renwick (Eds.), *Proceedings of the 7th International Conference on Music Perception and Cognition* (pp. 469–472). Adelaide: Causal Productions.

Aarden, B. (2003). *Dynamic melodic expectancy* (Doctoral dissertation). Ohio State University, Columbus. (UMI No. 3124024.)

Aarden, B., & von Hippel, P. T. (2004). Rules of chord doubling (and spacing): Which ones do we need? *Music Theory Online*, 10(2). http://mto.societymusictheory.org/issues/mto.04.10.2/mto.04.10.2.aarden_hippel_frames.html.

Aldwell, E., & Schachter, C. (1989). *Harmony and voice-leading* (2nd ed.). Orlando, FL: Harcourt, Brace, Jovanovich.

Aldwell, E., Schachter, C., & Cadwallader, A. (2010). *Harmony and voice leading* (4th ed.). New York: G. Schirmer.

ANSI. (1973). *Psychoacoustical terminology*. New York: American National Standards Institute.

Arthur, C., & Huron, D. (2016). The direct octaves rule: Testing a scene-analysis interpretation. *Musicae Scientiae*, 20(1), 1–17.

Attneave, F., & Olson, R. K. (1971). Pitch as a medium: A new approach to psychophysical scaling. *American Journal of Psychology*, 84, 147–166.

Beauvois, M. W. (1998). The effect of tone duration on auditory stream formation. *Perception and Psychophysics*, 60(5), 852–861.

Békésy, G. von. (1943/1949). Über die Resonanzkurve und die Abklingzeit der verschiedenen Stellen der Schneckentrennwand. *Akustiche Zeitschrift*, 8, 66–76.

Translation: On the resonance curve and the decay period at various points on the cochlear partition. *Journal of the Acoustical Society of America*, *1949*(21), 245–254.

Békésy, G. von. (1960). *Experiments in hearing*. New York: McGraw-Hill.

Benjamin, T. (1986). *Counterpoint in the style of J. S. Bach*. New York: Schirmer Books.

Benjamin, T., Horvit, M., & Nelson, R. (2007). *Techniques and materials of music: From the common practice period through the twentieth century* (7th ed.). New York: G. Schirmer.

Beranek, L. L. (1954/1986). *Acoustics*. New York: McGraw-Hill.

Berardi, A. (1681). *Documenti armonici*. Bologna, Italy.

Bharucha, J. J. (1984). Anchoring effects in music: The resolution of dissonance. *Cognitive Psychology*, *16*, 485–518.

Bodin, L. G. (1997). Aesthetic profile mapping—an outline of a practical aesthetic analysis. In *Proceedings of the International Academy of Electroacoustic Music* (Vol. 2, pp. 222–226).

Bonin, T. L., Trainor, L., Belyk, M., & Andrews, P. (under review). The source dilemma hypothesis: Perceptual uncertainty contributes to musical emotion.

Bozzi, P., & Vicario, G. (1960). Due fattori di unificazione fra note musicali: La vicinanza temporale e la vicinanza tonale. *Rivista di Psicologia*, *54*(4), 253–258.

Brandenberg, K. (1999). MP3 and AAC explained. In *Proceedings of the Audio Engineering Society 17th International Conference on High Quality Audio Coding* (pp. 1–12).

Bregman, A. S. (1978). Auditory streaming is cumulative. *Journal of Experimental Psychology: Human Perception and Performance*, *4*, 380–387.

Bregman, A. S. (1990). *Auditory scene analysis: The perceptual organization of sound*. Cambridge, MA: MIT Press.

Bregman, A. S., Abramson, J., Doehring, P., & Darwin, C. (1985). Spectral integration based on common amplitude modulation. *Perception and Psychophysics*, *37*, 483–493.

Bregman, A. S., Ahad, P. A., Crum, P. A. C., & O'Reilly, J. (2000). Effects of time intervals and tone durations on auditory stream segregation. *Perception and Psychophysics*, *62*(3), 626–636.

Bregman, A. S., & Campbell, J. (1971). Primary auditory stream segregation and perception of order in rapid sequences of tones. *Journal of Experimental Psychology*, *89*(2), 244–249.

Bregman, A. S., & Doehring, P. (1984). Fusion of simultaneous tonal glides: The role of parallelness and simple frequency relations. *Perception and Psychophysics*, *36*(3), 251–256.

Bregman, A. S., Liao, C., & Levitan, R. (1990). Auditory grouping based on fundamental frequency and formant peak frequency. *Canadian Journal of Psychology, 44,* 400–413.

Bregman, A. S., & Pinker, S. (1978). Auditory streaming and the building of timbre. *Canadian Journal of Psychology, 32*(1), 19–31.

Brunstrom, J. M., & Roberts, B. (1998). Profiling the perceptual suppression of partials in periodic complex tones: Further evidence for a harmonic template. *Journal of the Acoustical Society of America, 104*(6), 3511–3519.

Butler, D. (1979). A further study of melodic channeling. *Perception and Psychophysics, 25,* 264–268.

Callender, C. (1998). Voice-leading parsimony in the music of Alexander Scriabin. *Journal of Music Therapy, 42*(2), 219–233.

Cambouropoulos, E. (2008). Voice and stream: Perceptual and computational modeling of voice separation. *Music Perception, 26*(1), 75–94.

Cambouropoulos, E., & Tsougras, C. (2009). Auditory streams in Ligeti's *Continuum*: A theoretical and perceptual approach. *Journal of Interdisciplinary Music Studies, 3*(1–2), 119–137.

Carlsen, J. C. (1981). Some factors which influence melodic expectancy. *Psychomusicology: Music, Mind, and Brain, 1,* 12–29.

Carlyon, R. P. (1991). Discriminating between coherent and incoherent frequency modulation of complex tones. *Journal of the Acoustical Society of America, 89*(1), 329–340.

Carlyon, R. P. (1992). The psychophysics of concurrent sound segregation. In R. P. Carlyon, C. J. Darwin, & I. J. Russell (Eds.), *Processing of complex sounds by the auditory system* (pp. 53–61). Oxford: Clarendon Press/Oxford University Press.

Carlyon, R. P. (1994). Further evidence against an across-frequency mechanism specific to the detection of frequency modulation (FM) incoherence between resolved frequency components. *Journal of the Acoustical Society of America, 95*(2), 949–961.

Carter-Enyi, A. (2014, November 2). *Melodic languages and linguistic melodies: Singing in tone languages.* Paper presented at the College Music Society National Conference, St. Louis, MO.

Carter-Enyi, A., Aina, D., Carter-Enyi, Q., Huron, D., & Gomez, A. (2013, November 16). *Semblant motion in Nigerian praise music.* Paper presented at the Annual Meeting of the Society for Ethnomusicology, Indianapolis, IN.

Chon, S. H., & Huron, D. (2014). Does auditory masking explain high voice superiority? In *Proceedings of the International Conference on Music Perception and Cognition.* Seoul: Causal Productions.

Chon, S. H., & McAdams, S. (2012). Investigation of timbre saliency, the attention capturing quality of timbre. *Journal of the Acoustical Society of America, 131*(4), 3433.

Chorost, M. (2005). *Rebuilt: My journey back to the hearing world*. Boston: Houghton Mifflin.

Ciocca, V., & Bregman, A. S. (1987). Perceived continuity of gliding and steady-state tones through interrupting noise. *Perception and Psychophysics, 42*, 476–484.

Clendinning, J. P., & Marvin, E. W. (2010). *The musician's guide to theory and analysis* (2nd ed.). New York: W. W. Norton.

Cohn, R. (1996). Maximally smooth cycles, hexatonic systems, and the analysis of late-Romantic triadic progressions. *Music Analysis, 15*(1), 9–40.

Cohn, R. (1998). An introduction to neo-Riemannian theory: A survey and historical perspective. *Journal of Music Therapy, 42*(2), 167–180.

Cohn, R. (2012). *Audacious euphony: Chromaticism and the triad's second nature*. New York: Oxford University Press.

Crowder, R. G. (1969). Improved recall for digits with delayed recall cues. *Journal of Experimental Psychology, 82*, 258–262.

Cusack, R., Deeks, J., Aikman, G., & Carlyon, R. P. (2004). Effects of location, frequency region, and time course of selective attention on auditory scene analysis. *Journal of Experimental Psychology: Human Perception and Performance, 30*, 643–656.

Dadelsen, G. von. (1988). *Johann Sebastian Bachs Klavierbüchlein für Anna Magdalena Bach 1725*. Facsimile reproduction. Kassel: Bärenreiter-Verlag.

Danesi, M. (2002). *The puzzle instinct: The meaning of puzzles in human life*. Bloomington: Indiana University Press.

Darwin, C. J., & Gardner, R. B. (1986). Mistuning a harmonic of a vowel: Grouping and phase effects on vowel quality. *Journal of the Acoustical Society of America, 79*, 838–845.

Desain, P., Honing, H., & Sadakata (2003, June 18). *Predicting rhythm perception from rhythm production and score counts: The Bayesian approach*. Paper presented at the Society for Music Perception and Cognition 2003 Conference. Las Vegas, NV.

Descoeudres, A. (1921). *Le développement de l'enfant de deux à sept ans*. Paris: Delachaux & Niestlé.

Deutsch, D. (1975). Two-channel listening to musical scales. *Journal of the Acoustical Society of America, 57*, 1156–1160.

DeWitt, L. A., & Crowder, R. G. (1987). Tonal fusion of consonant musical intervals: The oomph in Stumpf. *Perception and Psychophysics, 41*(1), 73–84.

References

Divenyi, P. L., & Oliver, S. K. (1989). Resolution of steady-state sounds in simulated auditory space. *Journal of the Acoustical Society of America*, *85*, 2042–2052.

Dmytryk, E. (1984). *On film editing: An introduction to the art of film construction*. Abingdon, UK: Focal Press.

Dowling, W. J. (1967). *Rhythmic fission and the perceptual organization of tone sequences* (Doctoral dissertation). Harvard University, Cambridge, MA. (UMI No. 0219408.)

Dowling, W. J. (1973). The perception of interleaved melodies. *Cognitive Psychology*, *5*, 322–337.

Duane, B. (2008). *Auditory stream segregation and Schubert's piano sonata in B-flat, D. 960*. Presentation at the Society for Music Theory Annual meeting, Nashville, TN.

Duane, B. (2013). Auditory streaming cues in eighteenth- and early nineteenth-century string quartets: A corpus-based study. *Music Perception*, *31*(1), 46–58.

Duifhuis, H., Willems, L. F., & Sluyter, R. J. (1982). Measurement of pitch in speech: An implementation of Goldstein's theory of pitch perception. *Journal of the Acoustical Society of America*, *71*, 1568–1580.

Egan, J. P., & Hake, H. W. (1950). On the masking pattern of a simple auditory stimulus. *Journal of the Acoustical Society of America*, *22*(5), 622–630.

Ekwueme, L. (1974). Linguistic determinants of some Igbo musical properties. *Journal of African Studies*, *1*(3), 335–353.

Erickson, R. (1975). *Sound structure in music*. Berkeley: University of California Press.

Estes, B., & Combes, A. (1966). Perception of quantity. *Journal of Genetic Psychology*, *108*, 333–336.

Fétis, F.-J. (1840). *Esquisse de l'histoire de l'harmonie*. Paris: Revue et gazette musicale de Paris. English trans. by M. Arlin, Stuyvesant, NY: Pendragon Press, 1994.

Fitts, P. M. (1954). The information capacity of the human motor system in controlling amplitude of movement. *Journal of Experimental Psychology*, *47*, 381–391.

Fletcher, H. (1940). Auditory patterns. *Reviews of Modern Physics*, *12*, 47–65.

Fletcher, H. (1953). *Speech and hearing in communication*. New York: Van Nostrand.

Fux, J. J. (1725). *Gradus ad Parnassum*. Vienna.

Gauldin, R. (1985). *A practical approach to sixteenth-century counterpoint*. Englewood Cliffs, NJ: Prentice Hall.

Geer, J. P. Van de, Levelt, W. J. M., & Plomp, R. (1962). The connotation of musical consonance. *Acta Psychologica*, *20*, 308–319.

Gelman, R., & Tucker, M. F. (1975). Further investigations of the young child's conception of number. *Child Development, 46*, 167–175.

Gjerdingen, R. O. (1994). Apparent motion in music? *Music Perception, 11*(4), 335–370.

Gjerdingen, R. O. (2007). *Music in the Galant style*. Oxford: Oxford University Press.

Glasberg, B. R., & Moore, B. C. J. (1990). Derivation of auditory filter shapes from notched noise data. *Hearing Research, 47*, 103–138.

Glucksberg, S., & Cowen, G. N., Jr. (1970). Memory for nonattended auditory material. *Cognitive Psychology, 1*, 149–156.

Gregory, A. H. (1994). Timbre and auditory streaming. *Music Perception, 12*, 161–174.

Greenwood, D. D. (1961a). Auditory masking and the critical band. *Journal of the Acoustical Society of America, 33*(4), 484–502.

Greenwood, D. D. (1961b). Critical bandwidth and the frequency coordinates of the basilar membrane. *Journal of the Acoustical Society of America, 33*(4), 1344–1356.

Greenwood, D. D. (1990). A cochlear frequency-position function for several species—29 years later. *Journal of the Acoustical Society of America, 87*(6), 2592–2605.

Greenwood, D. D. (1991). Critical bandwidth and consonance in relation to cochlear frequency-position coordinates. *Hearing Research, 54*(2), 164–208.

Guttman, N., & Julesz, B. (1963). Lower limits of auditory periodicity analysis. *Journal of the Acoustical Society of America, 35*, 610.

Haas, H. (1951). Über den Einfluß eines Einfachechos auf die Hörsamkeit von Sprache. *Acustica, 1*, 49–58. Translated by K. P. R. Ehrenberg as The influence of a single echo on the audibility of speech. *Journal of the Audio Engineering Society, 20*, 145–159.

Hall, M. D., & Pastore, R. E. (1992). Musical duplex perception: Perception of figurally good chords with subliminal distinguishing tones. *Journal of Experimental Psychology: Human Perception and Performance, 18*(3), 752–762.

Hartmann, W. M., & Johnson, D. (1991). Stream segregation and peripheral channeling. *Music Perception, 9*(2), 155–183.

Heise, G. A., & Miller, G. A. (1951). An experimental study of auditory patterns. *American Journal of Psychology, 64*, 68–77.

Helmholtz, H. L. von. (1863/1878). *Die Lehre von der Tonempfindungen als physiologische Grundlage für die Theorie der Musik*. Braunschweig: Verlag Friedrich Vieweg & Sohn. Translated by Alexander Ellis as *On the sensations of tone as a physiological basis for the theory of music*. (1878); 2nd English ed. New York: Dover, 1954.

Hindemith, P. (1944). *A concentrated course in traditional harmony* (2nd ed., 2 vols.). New York: Associated Music Publishers.

Hippel, P. von. (2000). Redefining pitch proximity: Tessitura and mobility as constraints on melodic interval size. *Music Perception, 17*(3), 315–327.

Hippel, P. von, & Huron, D. (2000). Why do skips precede reversals? The effect of tessitura on melodic structure. *Music Perception, 18*(1), 59–85.

Hirsh, I. J. (1959). Auditory perception of temporal order. *Journal of the Acoustical Society of America, 31*(6), 759–767.

Horwood, F. J. (1948). *The basis of harmony*. Toronto: Gordon V. Thompson.

Houtgast, T. (1971). Psychophysical evidence for lateral inhibition in hearing. In *Proceedings of the 7th International Congress on Acoustics* (Vol. 3, pp. 521–524).

Houtgast, T. (1972). Psychophysical evidence for lateral inhibition in hearing. *Journal of the Acoustical Society of America, 51*, 1885–1894.

Houtgast, T. (1973). Psychophysical experiments on "tuning curves" and "two-tone inhibition." *Acustica, 29*, 168–179.

Hurley, B., Martens, P., & Janata, P. (2014). Spontaneous sensorimotor coupling with multipart music. *Journal of Experimental Psychology: Human Perception and Performance, 40*(4), 1679–1696.

Huron, D. (1987). *Auditory stream segregation and voice independence in multi-voice musical textures*. Paper presented at the Second Conference on Science and Music, City University, London, UK.

Huron, D. (1989a). *Voice segregation in selected polyphonic keyboard works by Johann Sebastian Bach* (Doctoral dissertation). University of Nottingham, Nottingham, England. (UMI No. U011957)

Huron, D. (1989b). Voice denumerability in polyphonic music of homogeneous timbres. *Music Perception, 6*(4), 361–382.

Huron, D. (1989c). Characterizing musical textures. In *Proceedings of the 1989 International Computer Music Conference* (pp. 131–134). San Francisco, CA: Computer Music Association.

Huron, D. (1990). Increment/decrement asymmetries in polyphonic sonorities. *Music Perception, 7*(4), 385–393.

Huron, D. (1991a). The avoidance of part-crossing in polyphonic music: Perceptual evidence and musical practice. *Music Perception, 9*(1), 93–104.

Huron, D. (1991b). Tonal consonance versus tonal fusion in polyphonic sonorities. *Music Perception, 9*(2), 135–154.

Huron, D. (1991c). Albert S. Bregman: Auditory scene analysis: The perceptual organization of sound [review]. *Psychology of Music, 19*(1), 77–82.

Huron, D. (1992). The ramp archetype and the maintenance of auditory attention. *Music Perception, 10*(1), 83–92.

Huron, D. (1993a). Note-onset asynchrony in J. S. Bach's two-part Inventions. *Music Perception, 10*(4), 435–443.

Huron, D. (1993b). Chordal-tone doubling and the enhancement of key perception. *Psychomusicology: Music, Mind, and Brain, 12*(1), 73–83.

Huron, D. (1993c). A derivation of the rules of voice-leading from perceptual principles. *Journal of the Acoustical Society of America, 93*(4), S2362.

Huron, D. (1994). Interval-class content in equally-tempered pitch-class sets: Common scales exhibit optimum tonal consonance. *Music Perception, 11*(3), 289–305.

Huron, D. (2001). Tone and voice: A derivation of the rules of voice-leading from perceptual principles. *Music Perception, 19*(1), 1–64.

Huron, D. (2005). The plural pleasures of music. In W. Brunson & J. Sundberg (Eds.), *Proceedings of the 2004 Music and Science Conference* (pp. 65–78). Stockholm: Kungliga Musikhögskolan Förlaget.

Huron, D. (2006a). On teaching voice leading from perceptual principles. *Music Theory Pedagogy, 20*, 163–166.

Huron, D. (2006b). *Sweet anticipation: Music and the psychology of expectation.* Cambridge, MA: MIT Press.

Huron, D. (2007). On the role of embellishment tones in the perceptual segregation of concurrent musical parts. *Empirical Musicology Review, 2*(4), 123–139.

Huron, D. (2008). Asynchronous preparation of tonally fused intervals in polyphonic music. *Empirical Musicology Review, 3*(1), 11–21.

Huron, D. (2013). A psychological approach to musical form: The habituation-fluency theory of repetition. *Current Musicology, 96*, 7–35.

Huron, D. (2015a). Affect induction through musical sounds: An ethological perspective. *Philosophical Transactions of the Royal Society of London, Series B, Biological Sciences, 370*(1664), 1–7.

Huron, D. (2015b). Aesthetics. In S. Hallam, I. Cross, & M. Thaut (Eds.), *Oxford Handbook of Music Psychology* (2nd ed., pp. 233–245). Oxford: Oxford University Press.

Huron, D., & Fantini, D. (1989). The avoidance of inner-voice entries: Perceptual evidence and musical practice. *Music Perception, 7*(1), 43–47.

Huron, D., & Margulis, E. H. (2010). Music, expectation and frisson. In P. Juslin & J. Sloboda (Eds.), *Handbook of music and emotion: Theory, research, applications* (2nd ed., pp. 575–604). Oxford: Oxford University Press.

Huron, D., & Mondor, T. (1994). *Melodic line, melodic motion, and Fitts's law*. Unpublished manuscript.

Huron, D., & Parncutt, R. (1992). *How "middle" is middle C? Terhardt's virtual pitch weight and the distribution of pitches in music*. Unpublished manuscript.

Huron, D., & Sellmer, P. (1992). Critical bands and the spelling of vertical sonorities. *Music Perception*, *10*(2), 129–149.

Iverson, P. (1992). Auditory stream segregation by musical timbre. Paper presented at the 123rd meeting of the Acoustical Society of America.

Iversen, J. R., Patel, A. D., & Ohgushi, K. (2008). Perception of rhythmic grouping depends on auditory experience. *Journal of the Acoustical Society of America*, *124*, 2263–2271.

Iyer, N., Aarden, B., Hoglund, E., & Huron, D. (1999). Effect of intensity on sensory dissonance. *Journal of the Acoustical Society of America*, *106*(4), 2208–2209.

Jeppesen, K. (1931/1939/1963). *Kontrapunkt (vokalpolyfoni)*. Copenhagen: Wilhelm Hansen, 1931. Translated 1939 from Danish edition by G. Haydon as *Counterpoint: The polyphonic vocal style of the sixteenth century*. Englewood Cliffs, NJ: Prentice Hall, 1963.

Johnson, R. B. (2010). *Selected topics in the perception and interpretation of musical tempo* (Doctoral dissertation). Ohio State University, Columbus. (UMI No. 3425288)

Kaestner, G. (1909). Untersuchungen über den Gefühlseindruck unanalysierter Zweierklänge. *Psychologische Studien*, *4*, 473–504.

Kameoka, A., & Kuriyagawa, M. (1969a). Consonance theory part I: Consonance of dyads. *Journal of the Acoustical Society of America*, *45*, 1451–1459.

Kameoka, A., & Kuriyagawa, M. (1969b). Consonance theory part II: Consonance of complex tones and its calculation method. *Journal of the Acoustical Society of America*, *45*, 1460–1469.

Keys, I. (1961). *The texture of music: From Purcell to Brahms*. London: Dennis Dobson.

Klumpenhouwer, H. (1994). Some remarks on the use of Riemann transformations. *Music Theory Online*, 0, 9.

Korte, A. (1915). Kinomatoscopische Untersuchungen. *Zeitschrift für Psychologie der Sinnesorgane*, *72*, 193–296.

Kostka, S., Payne, D., & Almén, B. (2012). *Tonal harmony* (7th ed.). New York: McGraw-Hill.

Krumhansl, C. (1979). The psychological representation of musical pitch in a tonal context. *Cognitive Psychology, 11*, 346–374.

Kubovy, M., & Howard, F. P. (1976). Persistence of pitch-segregating echoic memory. *Journal of Experimental Psychology: Human Perception and Performance, 2*(4), 531–537.

Laitz, S. G. (2012). *The complete musician: An integrated approach to tonal theory, analysis, and listening* (3rd ed.). New York: Oxford University Press.

Landy, L. (2007). *Understanding the art of sound organization.* Cambridge, MA: MIT Press.

Lerdahl, F., & Jackendoff, R. S. (1985). *A generative theory of tonal music.* Cambridge, MA: MIT Press.

Lewin, D. (1987). *Generalized musical intervals and transformations.* New Haven, CT: Yale University Press.

MacDougall, H. G., & Moore, S. T. (2005). Marching to the beat of the same drummer: The spontaneous tempo of human locomotion. *Journal of Applied Physiology, 99*, 1164–1173.

Marie, C., & Trainor, L. (2013). Development of simultaneous pitch encoding: Infants show a high voice superiority effect. *Cerebral Cortex, 23*, 660–669.

Martin, D. (1965). The evolution of barbershop harmony. In *Music Journal Annual* (p. 40).

McAdams, S. (1982). Spectral fusion and the creation of auditory images. In M. Clynes (Ed.), *Music, mind, and brain: The neuropsychology of music* (pp. 279–298). New York: Plenum Press.

McAdams, S. (1984). *Spectral fusion, spectral parsing and the formation of auditory images* (Doctoral dissertation). Stanford University, Palo Alto, CA. (UMI No. 8420589.)

McAdams, S. (1989). Segregation of concurrent sounds: Effects of frequency modulation coherence. *Journal of the Acoustical Society of America, 86*, 2148–2159.

McAdams, S., & Bregman, A. S. (1979). Hearing musical streams. *Computer Music Journal, 3*(4), 26–43, 60, 63.

McNabb, M. (1983). Dreamsong (1978). *Michael McNabb/Computer Music* [LP]. Berkeley: 1750 Arch Street Records.

Merriam, A. P., Whinery, S., & Fred, B. G. (1956). Songs of a Rada community in Trinidad. *Anthropos, 51*, 157–174.

Meyer, L. B. (1956). *Emotion and meaning in music.* Chicago: University of Chicago Press.

Mikulas, W. L. (1996). Sudden onset of subjective dimensionality: A case study. *Perceptual and Motor Skills, 82*(3), 852–854.

Miller, G. A., & Heise, G. A. (1950). The trill threshold. *Journal of the Acoustical Society of America, 22*(5), 637–638.

Moore, B. C. J., & Glasberg, B. R. (1983). Suggested formulae for calculating auditory-filter bandwidths and excitation patterns. *Journal of the Acoustical Society of America, 74*(3), 750–753.

Morris, R. O. (1946). *The Oxford harmony* (Vol. 1). London: Oxford University Press.

Mursell, J. L. (1937). *The psychology of music*. New York: W. W. Norton.

Narmour, E. (1990). *The analysis and cognition of basic melodic structures: The implication-realization model*. Chicago: University of Chicago Press.

Narmour, E. (1992). *The analysis and cognition of melodic complexity: The implication-realization model*. Chicago: University of Chicago Press.

Neisser, U. (1967). *Cognitive psychology*. New York: Meredith.

Noll, P. (1993). Wideband speech and audio coding. *IEEE Communications Magazine, 13*(11), 34–44.

Noorden, L. P. A. S. van. (1971a). Rhythmic fission as a function of tone rate. *IPO Annual Progress Report, 6*, 9–12.

Noorden, L. P. A. S. van. (1971b). Discrimination of time intervals bounded by tones of different frequencies. *IPO Annual Progress Report, 6*, 12–15.

Noorden, L. P. A. S. van. (1975). *Temporal coherence in the perception of tone sequences* (Doctoral dissertation). Technisch Hogeschool Eindhoven; published Eindhoven: Druk vam Voorschoten.

Norman, D. (1967). Temporal confusions and limited capacity processors. *Acta Psychologica, 27*, 293–297.

Ohgushi, K., & Hato, T. (1989). On the perception of the musical pitch of high frequency tones. In *Proceedings of the 13th International Congress on Acoustics* (Vol. 3, pp. 27–30).

Opolko, F., & Wapnick, J. (1989). *McGill University Master Samples. Compact discs* (Vol. 1–11). Montreal: McGill University, Faculty of Music.

Ortmann, O. R. (1926). *On the melodic relativity of tones*. Princeton, NJ: Psychological Review Company.

Parks, R. (1984). *Eighteenth-century counterpoint and tonal structure*. Englewood Cliffs, NJ: Prentice Hall.

Parncutt, R. (1989). *Harmony: A psychoacoustical approach*. Berlin: Springer-Verlag.

Parncutt, R. (1993). Pitch properties of chords of octave-spaced tones. *Contemporary Music Review*, *9*, 35–50.

Patel, A. D., Iversen, J. R., & Ohgushi, K. (2004). Cultural differences in rhythm perception: What is the influence of native language? In S. D. Lipscomb, R. Ashley, R. O. Gjerdingen, & P. Webster (Eds.), *Proceedings of the 8th International Conference on Music Perception and Cognition* (pp. 88–89). Evanston, IL: Casual Productions.

Patterson, R. D. (2000). Auditory images: How complex sounds are represented in the auditory system. *Journal of the Acoustical Society of Japan*, *21*(4), 183–190.

Patterson, R. D., Robinson, K., Holdsworth, J., McKeown, D., Zhang, C., & Allerhand, M. (1992). Complex sounds and auditory images. In Y. Cazals, L. Demany, & K. Horner (Eds.), *Auditory physiology and perception: Proceedings of the Ninth International Symposium on Hearing* (pp. 429–446). Oxford: Pergamon.

Pattison, H., Gardner, R. B., & Darwin, C. J. (1986). Effects of acoustical context on perceived vowel quality. *Journal of the Acoustical Society of America*, *80* (Supplement 1).

Phillips, T. K. E. (1953). *Yoruba music (African): Fusion of speech and music*. Johannesburg, South Africa: African Music Society.

Piston, W. (1962/1978). *Harmony* (4th ed., revised by Mark Devoto). New York: W. W. Norton.

Plomp, R., & Levelt, W. J. M. (1965). Tonal consonance and critical bandwidth. *Journal of the Acoustical Society of America*, *37*, 548–560.

Price, P. (1984). Carillon. In S. Sadie (Ed.), *New Grove dictionary of musical instruments* (Vol. 1, pp. 307–311). London: Macmillan.

Preisler, A. (1993). The influence of spectral composition of complex tones and of musical experience on the perceptibility of virtual pitch. *Perception and Psychophysics*, *54*(5), 589–603.

Pressnitzer, D., & Hupé, J.-M. (2005). *Is auditory streaming a bistable percept?* (pp. 1557–1561). Budapest: Forum Acusticum.

Pressnitzer, D., Suied, C., & Shamma, S. A. (2011). Auditory scene analysis: The sweet music of ambiguity. *Frontiers in Human Neuroscience*, *5*, 158.

Rameau, J.-P. (1722). *Traité de l'harmonie*. Paris: Jean Baptiste Christophe Ballard. Facsimile reprint, New York: Broude Bros., 1965.

Rasch, R. A. (1978). The perception of simultaneous notes such as in polyphonic music. *Acustica*, *40*, 21–33.

Rasch, R. A. (1979). Synchronization in performed ensemble music. *Acustica*, *43*, 121–131.

References

Rasch, R. A. (1981). *Aspects of the perception and performance of polyphonic music* (Unpublished doctoral dissertation). Utrecht, Netherlands: Elinkwijk BV.

Rasch, R. A. (1988). Timing and synchronization in ensemble performance. In J. Sloboda (Ed.), *Generative processes in music: The psychology of performance, improvisation, and composition* (pp. 70–90). Oxford: Clarendon Press.

Remez, R. E., Rubin, P. E., Pisoni, D. B., & Carrell, T. D. (1981). Speech perception without traditional speech cues. *Science, 212*, 947–949.

Riemann, H. (1903). *Grosse Kompositionslehre*. Berlin and Stuttgart: W. Spemann.

Roig-Francoli, M. (2010). *Harmony in context* (2nd ed.). New York: McGraw-Hill.

Root, D. L. (1980). Barbershop harmony. In S. Sadie (Ed.), *New Grove dictionary of music and musicians* (Vol. 2, p. 137). London: Macmillan.

Rössler, E. K. (1952). *Klangfunktion und Registrierung*. Kassel: Bärenreiter Verlag.

Rostron, A. B. (1974). Brief auditory storage: Some further observations. *Acta Psychologica, 38*, 471–482.

Rubin, E. (1915). *Synsoplevede Figurer; Studier i psykologisk Analyse*. Denmark.

Sacks, O. (2007). *Musicophilia: Tales of music and the brain*. New York: Knopf.

Saffran, J. R., Johnson, E. K., Aslin, R. N., & Newport, E. L. (1999). Statistical learning of tone sequences by human infants and adults. *Cognition, 70*, 27–52.

Saldanha, E. L., & Corso, J. F. (1964). Timbre cues and the identification of musical instruments. *Journal of the Acoustical Society of America, 36*, 2021–2026.

Sandell, G. J. (1991a). A library of orchestral instrument spectra. In *Proceedings of the 1991 International Computer Music Conference* (pp. 98–101).

Sandell, G. J. (1991b). *Concurrent timbres in orchestration: A perceptual study of factors determining "blend"* (Doctoral dissertation). Northwestern University, Evanston, IL. (UMI No. 9213551.)

Schaeffer, B., Eggleston, V. H., & Scott, J. L. (1974). Number development in young children. *Cognitive Psychology, 6*, 357–379.

Scheffers, M. T. M. (1983). Simulation of auditory analysis of pitch: An elaboration on the DWS pitch meter. *Journal of the Acoustical Society of America, 74*, 1716–1725.

Schellenberg, M. (2012). Does language determine music in tone languages? *Ethnomusicology, 56*(2), 266–278.

Schellenberg, G., Corrigall, K., Ladinig, O., & Huron, D. (2012). Changing the tune: Listeners like music that expresses a contrasting emotion. *Frontiers in Psychology, 3*(574), 1–9.

Schenker, H. (1906). *Neue musikalische Theorien und Phantasien* (Vol. 1): *Harmonielehre*. Stuttgart and Berlin: J. G. Cotta'sche Buchhandlung Nachfolger.

Schneider, P., Sluming, V., Roberts, N., Bleeck, S., & Rupp, A. (2005). Structural, functional and perceptual differences in the auditory cortex of musicians and non-musicians predict musical instrument preference. *Annals of the New York Academy of Sciences, 1060*, 387–394.

Schoeffler, M., Stöter, F. R., Bayerlein, H., Edler, B., & Herre, J. (2013). An experiment about estimating the number of instruments in polyphonic music: A comparison between Internet and laboratory results. In *Proceedings of the International Society for Music Information Retrieval* (pp. 1–6).

Schoenberg, A. (1911/1978). *Harmonielehre*. Vienna: Universal Edition. Translated by R.E. Carter as *Theory of harmony*. Berkeley: University of California Press.

Schouten, J. F. (1962). On the perception of sound and speech. In *Proceedings of the 4th International Congress on Acoustics* (Vol. 2, pp. 201–203).

Schubert, P. (1999). *Modal counterpoint, Renaissance style*. New York: Oxford University Press.

Schubert, P., & Neidhöfer, C. (2005). *Baroque counterpoint*. Upper Saddle River, NJ: Prentice Hall.

Schulte, M., Knief, A., Seither-Preisler, A., & Pantev, C. (2002). Different modes of pitch perception and learning-induced neuronal plasticity in the human auditory cortex. *Neural Plasticity, 9*(3), 161–175.

Schutz, M., Keeton, K., Huron, D., & G. Loewer. (2008). The happy xylophone: Instrument acoustics and articulation rate influence modality preference. *Empirical Musicology Review, 3*(3), 126–135.

Seither-Preisler, A., Johnson, L., Krumbholz, K., Nobbe, A., Patterson, R. D., Seither, S., et al. (2007). Tone sequences with conflicting fundamental pitch and timbre changes are heard differently by musicians and non-musicians. *Journal of Experimental Psychology: Human Perception and Performance, 33*(3), 743–751.

Semal, C., & Demany, L. (1990). The upper limit of "musical" pitch. *Music Perception, 8*(2), 165–175.

Shepard, R. N. (1981). Psychophysical complementarity. In M. Kubovy & J. Pomerantz (Eds.), *Perceptual organization* (pp. 279–341). Hillsdale, NJ: Erlbaum.

Smaragdis, P., Raj, B., & Shashanka, M. V. (2007). Supervised and semi-supervised separation of sounds from single-channel mixtures. In M. Davies, C. James, S. Abdallah, & M. Plumbley (Eds.), *Independent components analysis and signal separation* (pp. 414–421). Berlin: Springer.

Smaragdis, P., Shashanka, M., & Raj, B. (2009). A sparse non-parametric approach for single channel separation of known sounds. In Y. Bengio, D. Schuurmans, J. Lafferty, C. Williams, & A. Culotta (Eds.), *Advances in neural information processing systems, 22* (pp. 1705–1713). Cambridge, MA: MIT Press.

Stainer, J. (1878). *Harmony*. London: Novello.

Steiger, H., & Bregman, A. S. (1981). Capturing frequency components of glided tones: Frequency separation, orientation and alignment. *Perception and Psychophysics, 30*, 425–435.

Strauss, M. S., & Curtis, L. E. (1981). Infant perception of numerosity. *Child Development, 52*(4), 1146–1152.

Strong, W., & Clark, M. (1967). Synthesis of wind-instrument tones. *Journal of the Acoustical Society of America, 41*(1), 39–52.

Stumpf, C. (1890). *Tonpsychologie* (Vol. 2). Leipzig: S. Hirzel.

Sundberg, J. (1987). *The science of the singing voice*. De Kalb: Northern Illinois University Press.

Sundberg, J., Askenfelt, A., & Frydén, L. (1983). Music performance: A synthesis-by-rule approach. *Computer Music Journal, 7*(1), 37–43.

Tatsuoka, M., & Tiedeman, D. (1954). Discriminant analysis. *Review of Educational Research, 24*(5), 402–420.

Temperley, D. (2004). *The cognition of basic musical structures*. Cambridge, MA: MIT Press.

Temperley, D. (2007). *Uniform information density in music*. Paper presented at the 2007 Society for Music Perception Conference, Montreal.

Temperley, D. (2014). Information flow and repetition in music. *Journal of Music Therapy, 58*(2), 155–178.

Terhardt, E., Stoll, G., & Seewann, M. (1982a). Pitch of complex signals according to virtual-pitch theory: Tests, examples, and predictions. *Journal of the Acoustical Society of America, 71*(3), 671–678.

Terhardt, E., Stoll, G., & Seewann, M. (1982b). Algorithm for extraction of pitch and pitch salience from complex tonal signals. *Journal of the Acoustical Society of America, 71*(3), 679–688.

Terhardt, E., Stoll, G., Schermbach, R., & Parncutt, R. (1986). Tonhöhenmehrdeutigkeit, Tonverwandtschaft und Identifikation von Sukzessivintervallen. *Acustica, 61*, 57–66.

Thompson, W. F., & Schellenberg, E. G. (2006). Listening to music. In *MENC handbook of musical cognition and development* (pp. 72–123). Chicago: Music Educators National Conference.

Thurlow, W. R. (1957). An auditory figure-ground effect. *American Journal of Psychology, 70*, 653–654.

Todd, N. (1995). The kinematics of musical expression. *Journal of the Acoustical Society of America, 97*(3), 1940–1949.

Todd, N., Cousins, R., & Lee, C. S. (2007). The contribution of anthropometric factors to individual differences in the perception of rhythm. *Empirical Musicology Review, 2*(1), 1–13.

Tougas, Y., & Bregman, A. S. (1985). Crossing of auditory streams. *Journal of Experimental Psychology: Human Perception and Performance, 11*(6), 788–798.

Trainor, L. J., Marie, C., Bruce, I. C., & Bidelman, G. M. (2014). Explaining the high voice superiority effect in polyphonic music: Evidence from cortical evoked potentials and peripheral auditory models. *Hearing Research, 308*, 60–70.

Treisman, A. M. (1964). Monitoring and storage of irrelevant messages in selective attention. *Journal of Verbal Learning and Verbal Behavior, 3*, 449–459.

Treisman, M., & Howarth, C. I. (1959). Changes in threshold level produced by a signal preceding or following the threshold stimulus. *Quarterly Journal of Experimental Psychology, 11*, 129–142.

Treisman, M., & Rostron, A. B. (1972). Brief auditory storage: A modification of Sperling's paradigm applied to audition. *Acta Psychologica, 36*, 161–170.

Trythall, G. (1993). *Eighteenth century counterpoint*. Madison, WI: Brown & Benchmark.

Tymoczko, D. (2008). Scale theory, serial theory, and voice leading. *Music Analysis, 27*(1), 1–49.

Tymoczko, D. (2011). *A geometry of music*. New York: Oxford University Press.

Veeksman, K., Ressel, L., Mueller, J., Vischer, M., & Brockmeier, S. J. (2009). Comparison of music perception in bilateral and unilateral cochlear implant users and normal-hearing subjects. *Audiology and Neuro-Otology, 14*(5), 315–326.

Vicario, G. (1960). Analisi sperimentale di un caso di dipendenza fenomenica tra eventi sonori. *Rivista di Psicologia, 54*(3), 83–106.

Vos, J. (1995). Perceptual separation of simultaneous complex tones: The effect of slightly asynchronous onsets. *Acta Acustica (Les Ulis), 3*, 405–416.

Warren, R. M., Obusek, C. J., & Ackroff, J. (1972). Auditory induction: Perceptual synthesis of absent sounds. *Science, 176*, 1149–1151.

References

Wessel, D. L. (1979). Timbre space as a musical control structure. *Computer Music Journal*, *3*(2), 45–52.

Woodrow, H. (1909). A quantitative study of rhythm: The effect of variations in intensity, rate and duration. *Archives de Psychologie*, *14*, 1–166.

Wright, J. K. (1986). *Auditory object perception: Counterpoint in a new context* (Unpublished master's thesis). McGill University, Montreal, Canada.

Wright, J. K., & Bregman, A. S. (1987). Auditory stream segregation and the control of dissonance in polyphonic music. *Contemporary Music Review*, *2*, 63–93.

Yeary, M. J. (2011). *Perception, pitch, and musical chords* (Doctoral dissertation). University of Chicago, Chicago, IL. (UMI No. 3472979.)

Index

Page numbers joined by a "+" indicate that both pages are needed for a complete statement of the principle. For a more extensive index, refer to the book's website.

Aarden, Bret, 131, 135, 151, 153–156, 221, 226n12, 232n13, 232n4, 232n8, 233n1, 233n2, 233n4
Acoustic
 phenomena, 23–24
 scene, 26
Acoustic/auditory distinction, 23–24
"Adeste Fideles," 101–105
Alberti bass, 173
Aldwell, Edward, 158, 223n1, 223n2, 231n1, 233n6
"All Hail the Power of Jesus' Name," 203–204
Almén, Byron, 157, 230n23, 233n3
Analytic listening, 31, 132–133, 167–168, 173, 210. *See also* Synthetic listening
Andrews, Paul, 209, 235n1, 236n2, 236n4
Anticipation
 embellishment, 121, 123, 125
 psychological, 130, 132
Aperiodic sounds, 27
Architectural acoustics, 200
Arthur, Claire, 158–161, 221, 233n7
Askenfelt, Anders, 80, 228n24
Asynchronous onsets. *See* Onset asynchrony

Attention, 116, 127, 170, 208
Auditory/acoustic distinction, 23–24
Auditory image, 16, 23–24, 66
Auditory induction, 64–66
Auditory scene analysis, 4, 18, 22–23, 25, 129. *See also* Principles of auditory scene analysis
Auditory stream, 24–25, 70, 209. See also Musical stream; Textural stream
Augmented intervals rule, 12, 145, 213
Autophase method, 99–102
Avoid uncommon intervals rule, 145, 213

Bach, Anna Magdalena, 9
Bach, Johann Christoph Friedrich, 9
Bach, Johann Sebastian, 9, 45, 57, 59–60, 62, 77, 84, 92, 100, 102–105, 107–108, 111, 122–127, 151, 154, 156, 226n15, 226n16, 226n20, 228n16
 Die Kunst der Fuge, 111
 Das musikalische Opfer, 111
 organ fugues, 111
 Sinfonia No. 1 (BWV 787), 101–105
 solo violin fugues, 111
 two-part *Inventions*, 100, 111

Bach, Johann Sebastian (cont.)
  three-part *Sinfonias*, 101, 111
  Well-Tempered Clavier, fugues, 111
Backward music, 133–135
Barbershop style, 5, 51, 117, 214
Baroque part-writing canon
  augmented intervals rule, 12
  avoid leaps rule, 11
  chord spacing rule, 10
  common tone rule, 11, 211
  compass rule, 10, 211
  consecutive perfects rule, 11
  direct (or hidden or exposed) octaves (and fifths) rule, 11
  false (or cross-) relation rule, 12
  nearest tone rule, 11
  parallel perfects rule, 11
  part crossing rule, 11
  part overlap rule, 11
  step motion rule, 11
  tendency doubling rule, 11
  textural density rule, 10
  unisons rule, 11, 91, 211
Basilar membrane, 15, 41, 152, 187. *See also* Critical bands
  and chord spacing, 45–46
  length, 42
  map, 43–45
Békésy, Georg von, 15, 224n3, 225n2
Benjamin, Thomas, 157, 223n2, 233n2
Beranek, Leo, 229n4
Berardi, Angelo, 223n1
Bharucha, Jamshed, 231n5
Bolero (Maurice Ravel), 85, 177
Bonin, Tanor, 209, 235n1, 236n2, 236n4
Bottom-up, 129, 146, 191–192. *See also* Top-down
Brahms, Johannes, 215
Brain plasticity, 186, 188, 191, 193, 217
Bregman, Albert, 19, 59–60, 82, 92, 129, 221, 223n2, 223n4, 223n5, 223n5, 226n17, 227n10, 227n4, 227n8, 228n14, 229n1, 229n27, 229n8, 230n25, 231n5, 232n1, 224n8
Butler, David, 221, 227n13

Cadential 6–4, 140, 156
Cadwallader, Allen, 158, 233n6
Callender, Clifton, 223n1
Cambouropoulos, Emilios, 164, 234n2, 234n3
Campbell, Jeffrey, 25, 223n5, 224n8, 227n10, 227n8
Car horn, 163–164, 185, 192
Cariani, Peter, 221
Carlsen, James, 232n8
Carlyon, Robert, 166, 170, 223n5, 234n5, 236n6
Carter-Enyi, Aaron, 203, 236n17, 236n19
Causality vs. correlation, 149
Chon, Song Hui, 231n29
Chorost, Michael, 188, 235n6. *See also* Cochlear implants
Chromatic backtracking rule, 144, 213
Chromatic resolution rule, 144, 213
Clarity, 34, 37–38
Clendinning, Jane, 157, 233n5
Cochlea, 15, 41
Cochlear implants, 187–188, 191, 201
Cohn, Richard, 223n1
Common tone rule, 11, 91, 211
Compass rule, 10, 89, 211
Complex tone, 14
Compound fifths rule, 91, 211
Compound melodic line, 70
Condit-Schultz, Nat, 221
Conjunct motion, 72–73, 91, 212
Conjunct preparation rule, 94, 212
Continuity, 64, 67, 86, 207. *See also* Memory
Contrapuntal motions
  contrary, 82
  oblique, 82
  oblique preparation rule, 212

nonsemblant, 83
nonsemblant preparation rule, 212
parallel, 22, 24, 82–84, 169, 219
parallel motion rule, 212
semblant, 83, 123, 203
semblant motion rule, 212
similar, 82–83
Critical band, 42–43 45, 47, 53, 56, 152. *See also* Basilar membrane
Crossing parts, 74–77, 92, 212
Cuddy, Lola, 135, 221
Cusack, Rhodri, 166, 170, 234n5, 236n6

"Dance of the Reed Pipes" from *The Nutcracker* (Pyotr Tchaikovsky), 164, 167, 174–175, 177
Darwin, Chris, 228n14, 229n1
Debussy, Claude, 169
  La cathédrale engloutie (The Sunken Cathedral), 169, 178
Desain, Peter, 232n13
Descartes, René, 224n7. *See also* Theater of the mind
Descoeudres, Alice, 110, 230n19
Deutsch, Dianna, 75–76, 227n13
Direct intervals, 2, 5, 134, 157–162, 212
Dissonance, 53–62, 122
Divenyi, Pierre, 117, 231n31
Doubling rule, 156, 213
Dowling, Jay, 72, 131–133, 142, 221, 227n10, 227n11, 227n12, 227n13, 227n8, 232n5, 232n7
Dreamsong (Michael McNabb), 176–177, 179
Dropout, 178, 182, 196
Duane, Ben, 168, 202, 234n7, 234n8, 234n9, 236n14
Dvořák, Antonín, 45
Dysharmonia, 170

Echoes, 98, 200. *See also* Haas effect; Reverberation
Eitan, Zohar, 221

Embellishment rules, 124–125, 127, 213
Emotion
  beauty, viii
  emotional goals, 6, 104
  hedonic goals, 203
  irritation, 21, 55–56
  pleasure, 3, 6, 20–22, 55–56, 59, 196–198, 200–202, 204, 208–209
  plural pleasures, 202, 205
  relief, 21
  rewards, 21–22, 197
  scene density and pleasure, 197
English dissonance, 12
Environment's effect on hearing, viii, 185–193
Erickson, Robert, 231n28
Expectation, 130–145, 208
  dynamic-adaptive expectations, 141–142
  schematic expectations, 135–139, 142, 144–145
  unconscious expectation, 133
  veridical expectations, 135–136, 141–142

False relations, 12, 144–145, 213, 216
Fétis, François-Joseph, 223n1
Fitts, Paul, 77–78, 228n17, 228n20
Fitts's law, 78–82, 86. *See also* Leap lengthening
Fletcher, Harvey, 42, 47, 225n4
Follow tendencies rule, 144, 213
Frequency, 13
  fundamental, 14
"Frère Jacques," 131–132
Frustrated leading tone, 112, 160–161
Fryden, Lars, 80, 228n24
Fundamental pitch listeners, 186–187. *See also* Spectral pitch listeners
Fux, Johann, 223n1

Gabrieli, Giovanni, 117
Gauldin, Robert, 223n2

Geer, John van de, 55, 226n13
"Getting" polyphony, 216
Gjerdingen, Robert, 144, 164, 221, 228n21, 233n18, 233n2
Greenwood, Donald, 44, 225n5, 226n11, 226n12
Group final lengthening, 191
Guitar, 69, 104, 173–174 216

Haas, Helmut, 231n32, 235n8
Haas effect, 201
Hanon (Charles-Louis) piano exercise, 165, 167–168
"Happy Birthday," 129
Harmonic, 14
  fusion, 29–32, 59–60, 91, 123, 157–158, 160–161, 207
  series, 14, 44, 54
  sieve, 29–30, 186
  sounds, 27
  stick out, 30
  stretched harmonics, 29
Harmonicity, 27–29
Harmony deafness. *See* Dysharmonia
Haydn, Franz Joseph, 46, 80, 151, 154, 156
Hearing, 15
  apex (of cochlea), 15, 41
  base (of cochlea), 15, 41
  eardrum, 15
  equivalent rectangular bandwidth-rate, 43
  fundamental processing of pitch, 186–187
  middle of human hearing, 36
  ossicles, 15
  range of hearing, 42
  resolved partials, 16
  stapes, 15
Hearing out, 132, 158
Heise, George, 69–70, 72, 227n11, 227n7, 228n19
Helmholtz, Hermann von, 82, 229n26

Hertz, 13
High-voice superiority effect, 51, 127, 159, 215
Hindemith, Paul, 110, 223n1, 230n20, 231n1
Hippel, Paul von, 151, 153–156, 221, 228n23, 233n2, 233n4
Honing, Henkjan, 221, 232n13
Hurley, Brian, 197, 209, 235n12, 235n2, 236n3

Imitative part writing, 141
Implicit learning, 136–139, 192
Improvisation, 199, 209–210
Inharmonic sounds, 14, 27
Instrumentation balance, 116
Intelligibility of lyrics, 104–105, 125, 209
Interleaved melodies, 131–132, 142
Iversen, John, 191, 235n11, 235n12, 231n27
Ives, Charles, 234n8

Jackendoff, Ray, 229n1, 236n7
Janata, Petr, 197, 209, 235n12, 235n2, 236n3
Jeppesen, Knud, 223n2
Johnson, Randolph, 69, 227n6, 229n3

Kaestner, Georg, 56, 57, 60, 226n15
Kameoka, Akio, 54, 226n12, 226n14
Keys, Ivor, 223n1
Klumpenhouwer, Henry, 223n1
Korte, Adolf, 78, 228n18, 228n19, 228n20
Kostka, Stefan, 157, 230n23, 233n3
Krumhansl, Carol, 231n5
Kuriyagawa, Mamoru, 54, 226n12, 226n14

La cathédrale engloutie (Claude Debussy), 169, 178
Laitz, Steven, 157, 223n1, 233n1

Landy, Leigh, 178, 234n3
Leading tone, 135, 139–140
Leap away rule, 93, 125, 212
Leap lengthening, 80, 81, 92, 212, 214.
  See also Fitts's law
Lerdahl, Fred, 87, 229n1, 236n7
Levelt, Willem, 44–45, 54–55, 226n11, 226n13, 226n14, 229n4
Lewin, David, 223n1, 223n1
Ligeti, György, 234n2
Localization, 21, 22
  performers dispersed throughout an audience, 118
  source location, 19, 117–119, 208
  source location principle, statement of, 117
  spatial separation, 118

"Magic eye" pictures, 198
Magnet effect, 92
Margulis, Elizabeth, 135, 221, 232n8
Martens, Peter, 197, 209, 235n12, 235n2, 236n3
Marvin, Elizabeth, 157, 221, 233n5
Masking, 4, 41–54, 65
  evoked irritation, 53
  masker, 47
  minimum masking, 46, 207
  minimum masking principle, statement of, 47+52
  partial, 48–50, 53
  role of basilar membrane, 43–45, 49
  skirt, 48
McAdams, Stephen, 82, 221, 223n2, 223n5, 229n27, 229n28, 229n29, 230n25, 231n29
McNabb, Michael, 176–177, 179, 234n1
  Dreamsong, 176–177, 179
Meandering quality, 135, 141–142. See also Tending
Melismatic text setting, 203
Melody in highest voice. See High-voice superiority

Memory. See also continuity
  echoic, 66
  half-life of, 66
  short-term, 141
Merriam, Alan, 227n12
Meyer, Leonard, 122, 231n2
Micromodulation, 83, 215. See also Vibrato
Middle C, 34–37, 216. See also Toneness
Miller, George, 69–70, 72, 227n11, 227n7, 228n19
Missing fundamental, 32. See also Pitch
Mistuned harmonics, 29–30
Mode
  of excitation, 67
  of vibration, 13, 207
Motion
  apparent, 78
  conjunct, 72–73
  disjunct, 72–73
  Korte's third law of, 78
  leap, 73
  melodic, 77–80, 82
  muscles, 78, 79, 80, 81, 86
  step, 73, 159–160
Mozart, Wolfgang Amadeus, 80, 151, 154, 156
Muddled middle, 160
Mursell, James, 230n20
Musical stream, 170. See also Auditory stream; Textural stream
Musical texture, 101, 178
"My Bonnie Lies Over the Ocean," 80

Narmour, Eugene, 135, 221, 231n3, 232n8
Neidhöfer, Christoph, 223n2
Noorden, Leon van, 66, 71, 78, 80–81, 221, 223n2, 227n11, 227n13, 227n4, 227n8, 227n9, 228n19, 230n25
Norman, Donald, 227n8
The Nutcracker. See "Dance of the Reed Pipes"

Octaves rule, 211
"Oh Come All Ye Faithful," 101–105
Ohgushi, Kengo, 191, 224n6, 235n11, 235n12
Ollen, Joy, 221
"One, two, three, many" phenomenon, 110
Onset asynchrony, 97–108, 122, 168, 207
  asynchronous onsets rule (PR24), 106, 212
  asynchronous preparation rule (PR25), 106, 213
  measurement of, 98
  onset times, 97–98
  statement of principle, 98
Ortmann, Otto, 227n12
Out of sound, out of mind, 126
Overlap rule, 212

Pantev, Christo, 234n2, 234n4
Parncutt, Richard, 35, 109, 221, 223n1, 225n10, 225n9, 230n15, 230n16
Partial, 13
  resolved partial, 27–29
Partimenti, 144
Passing tone, 121, 123, 125
Patel, Aniruddh, 191, 235n11, 235n12
Patterson, Roy, 223n5, 234n2, 234n3
Payne, Dorothy, 157, 230n23, 233n3
Pedal tone, 121, 123, 125
Perfect parallels rule, 212
Petrouchka (Igor Stravinsky), 164, 167–168, 170, 177
Pinker, Steven, 223n5, 229n1
Piston, Walter, 46, 223n1, 223n2, 226n7, 231n1
Pitch, 34, 37–39
  ambiguity, 34
  as auditory phenomenon, 38–39
  average soprano, 45–46
  average alto, 45–46
  average tenor, 45–46

co-modulation, 82–83, 86, 123, 157, 161, 207
contour, 121
intonation, 78
learning, 186–187
pitch-class doubling rule, 125, 213
pitch-class duplication, 123–124
as prelinguistic sound label, 39
proximity, 72–82, 86, 122, 157, 161, 207
proximity and interleaved melodies, 142
successions, probabilities of, 137–140
Pleasure, 3, 6, 20–22, 55–56, 59, 196–198, 200–202, 204, 208–209
Plomp, Reinier, 44–45, 54–55, 226n11, 226n13, 226n14, 229n4
Plural pleasures, 202, 205
Praetorius, Michael, 99
Preference rules, 87, 219
  common tone rule (PR10), 91
  compass rule (PR2), 89
  compound fifths rule (PR8), 91
  conjunct motion rule (PR11), 91
  conjunct preparation rule (PR20), 94
  crossing rule (PR14), 92
  direct intervals rule (PR23), 95
  harmonic fusion rule (PR9), 91
  leap away rule (PR16), 93
  leap lengthening rule (PR12), 92
  nonsemblant preparation rule (PR21), 94
  oblique preparation rule (PR19), 94
  octaves rules (PR7), 91
  overlapping rule (PR15), 92
  parallel motion rule (PR18), 93
  perfect parallels rule (PR22), 95
  proximity rule (PR13), 92
  resist leaps rule (C3), 92
  semblant motion rule (PR17), 93
  spacing rule (PR4), 90
  sustained sound rule (PR3), 90
  tessitura rule (PR5), 90

Index

toneness rule (PR1), 89
unisons rules (PR6), 91
Prejudice against women, 51, 215
Prepared harmonic intervals, 106–107
Prepared perfect intervals, 107
Principles of auditory scene analysis, 207
　attention principle, statement of, 127
　continuity principle, statement of, 67
　expectation principle, statement of, 142
　harmonic fusion principle, statement of, 32
　limited density principle, statement of, 110
　minimum masking principle, statement of, 47+52
　onset asynchrony principle, statement of, 98
　pitch co-modulation principle, statement of, 84
　pitch proximity principle, statement of, 73+77+81
　timbral differentiation principle, statement of, 115
　toneness principle, statement of, 37
　source location principle, statement of, 117
Proximity rule, 212
Pseudo-polyphony, 70, 72, 111
Pure tone, 14
Pure tones, rarity of, 28
Puzzle
　jigsaw, 25–26, 129, 146
　pieces, 195, 197
　solving, 25, 129
Pyramid rule, 180

Rachael Y., 169–170
Rameau, Jean-Philippe, 223n1
Random-dot stereograms, 198
Rasch, Rudolph, 98–99, 221, 230n5, 230n6, 230n7

Ravel, Maurice, 85–86, 177
　Bolero, 85, 177
Reality vs. appearance, 16, 20, 23–24. *See also* Theater of the mind
Repetition, 122–123, 125, 146, 213
Retrograde music, 133–135. *See also* Time's arrow
Reverberation, 200–201, 205, 215. *See also* Echoes; Haas effect
Revised voice-leading canon, 210–214
Rhythmic energy, 105, 125
Rhythmic grouping, 190
Riemann, Hugo, 223n1
Rogers, Nancy, 221
Roig-Francoli, Miguel, 157, 223n1, 233n4
Rössler, Ernst, 231n28
"Row, Row, Row Your Boat," 142
Rubin, Edgar, 190, 235n8
Rubin, Philip, 189, 235n7

Sacks, Oliver, 169, 234n10, 234n11, 234n12
Sadakata, Makiko, 232n13
Saffran, Jenny, 232n12, 232n13
Sandell, Gregory, 116, 225n9, 231n30
Scale degree school, 150
Scene analysis trees, 5, 168, 173–174, 176, 182
Scene hierarchies, 173–174, 176, 178
Scene setting, 5, 178, 209–210
Schachter, Carl, 158, 223n1, 223n2, 231n1, 233n6
Scheffers, Michaël, 29, 224n1
Schellenberg, Glenn, 135, 234n6, 236n18, 236n5
Schemata, 144–145
Schenker, Heinrich, 223n1
Schmuckler, Mark, 135
Schneider, Peter, 186–187
Schoenberg, Arnold, 223n1
Schubert, Peter, 180, 221, 223n2, 223n2, 234n4

Schutz, Michael, 229n3
Second inversion chords, 153
Seither-Preisler, Annemarie, 186, 234n2, 234n3, 234n4
Shepard, Roger, 228n19, 230n16
Sinewave speech, 189, 191
Smoothness, 55, 59–61
"Somewhere Over the Rainbow" (Harold Arlen and E. Y. Harburg), 80
Sound, 13
Sounding as one, 59–61
Spacing rule, 211
Spectral
  envelope, 186
  pitch listeners, 186–187 (see also Fundamental pitch listeners)
  pitch processing, 186–187
  roll-off, 49
Statistical learning. See Implicit learning
Stereograms, 198
"St. Flavian," 134
Sustained sound rule, 211
Stereo, 19–22, 196, 201, 205, 215
Stimmführung, 4
Stöter, Fabian-Robert, 230n16
Stravinsky, Igor, 164, 177
  Petrouchka, 164, 167–168, 170, 177
Stream(s)
  auditory, 24–25, 70, 209
  contrasts, 178
  density, 111
  entries/exits, 178
  evolutions, 178
  fragmentations, 178
  hierarchical, 163–171
  mergers, 178
  musical, 170
Streaming, 73
Stride bass, 173–176
Stumpf, Carl, 59, 224n4
Sundberg, Johan, 80, 228n24, 229n25
The Sunken Cathedral (Claude Debussy), 169, 178

Synthetic listening, 173, 167–168, 210. *See also* Analytic listening

Tan, Daphne, 221
Tchaikovsky, Pyotr, 164, 167, 177
  "Dance of the Reed Pipes" from *The Nutcracker*, 164, 167, 174–175, 177
Temperely, David, 80, 87, 221, 228n23, 229n1, 229n7, 236n7
Tempo, 174–175, 214
Tendency tones, 139–140, 155
Tending, 135, 139–140. *See also* Meandering
Terhardt, Ernst, 34–35, 186, 224n7, 224n8, 225n9
Terretektorh (Iannis Xenakis), 118
Tessitura rule, 211
Textural stream, 168, 170, 209. See also Auditory stream; Musical stream
  inner voices as single textural stream, 160
  integrating musical parts into textural streams, 179
Texture, 176, 201, 210
  accompaniment, 175
  classification of, 101
  heterophony, 103
  homophony, 101–103, 210
  incremental, 197, 202
  melody, 175
  monophony, 103
  musical, 101, 178
  polyphonic-like homophony, 105, 114
  polyphony, 59, 101–103, 210
Theater of the mind, 23–24. *See also* Reality vs. appearance
Thomas, Dylan, 177
Thompson, William, 135, 234n6
Tillmann, Barbara, 135
Timbre, 114–116, 199, 207
Time's arrow, 133–135
Todd, Neil, 228n21

Tonal fusion rule, 211. *See also* Harmonic fusion
Toneness, 33–38, 207. *See also* Middle C
  maximum, 35–36
  tonality (psychoacoustic), 33
  toneness rule (PR1), 89, 211
Top-down, 129–130, 132–133, 142, 146, 191–192. *See also* Bottom-up
Trainor, Laurel, 209, 226n9, 235n1, 236n2, 236n4
Triad member school, 150
Trill boundary, 71, 86. *See also* Yodel boundary
Turn taking, vii, 125–127
"Twinkle, Twinkle, Little Star," 131–132
Tymoczko, Dmitri, 223n1

Un, deux, trois, beaucoup. *See* "One, two, three, many" phenomenon
Unisons rule, 11, 91, 211
Universal vs. innate, 185, 191

Vibrato, 83, 215. *See also* Micromodulation
Virtual instrument, 22, 24, 84–85
Virtual sound sources, 20, 118
Vision, 196
  annoyance, 52
  fear of the dark, 196
  glare, 196
  obstruction, 52, 196
  out of focus, 196
Voice leading, vii, 1, 135, 142, 155, 183, 193, 214
  definition of, 3, 135
  goal of, 88
  vs. part-writing, 135, 144, 146
  predictability of, 144
Vos, Joos, 106, 230n11

Wapnick, Joel, 225n9
Wessel, David, 114–115, 221, 230n25, 231n26

Wessel's illusion, 114–115
Why the Baroque canon?, 179–183
Women, prejudice against, 51, 215
Woodrow, Herbert, 190, 235n10
Wright, James, 122, 221, 223n2, 231n5, 232n6

Xenakis, Iannis, 118
  Terretektorh, 118

Yearning, 140
Yeary, Mark, 167, 221, 234n6
Yodel boundary, 71–72, 77–78, 80–81, 86. *See also* Trill boundary
Yodeling, 70, 72, 81, 173–174, 201, 216